SOLAR AND WIND ENERGY

SOLAR AND WIND ENERGY

An Economic Evaluation of Current and Future Technologies

MARTIN T. KATZMAN

Rowman & Allanheld
PUBLISHERS

ROWMAN & ALLANHELD

Published in the United States of America in 1984
by Rowman & Allanheld, Publishers
(A division of Littlefield, Adams & Company)
81 Adams Drive, Totowa, New Jersey 07512

Copyright © 1984 by Martin T. Katzman

All rights reserved. No part of this publication may
be reproduced, stored in a retrieval system, or
transmitted in any form or by any means, electronic,
mechanical, photocopying, recording, or otherwise,
without the prior permission of the publisher.

Library of Congress Cataloging in Publication Data

Katzman, Martin T.
  Solar and wind energy.

  Bibliography: p.
  Includes index.
  1. Solar energy—Economic aspects. 2. Wind power—
Economic aspects. I. Title.
TJ810.K28   1984   333.79'23   83-23044
ISBN 0-86598-152-3

84 85 86 / 10 9 8 7 6 5 4 3 2 1

Printed in the United States of America

To the next generation:
Doug, Karen, Sarah, and Julie

Contents

Figures	ix
Tables	xi
Preface	xiii
Acknowledgments	xix
Conversion Factors	xxi
Abbreviations	xxiii

1 Renewable Energy Sources and the Future   1
    The Dimensions of the Energy Problem, 1; Solar Energy as
    Manna, 5; The Menu of Technologies, 6; Is Solar Energy a
    Panacea?, 15; Notes, 16

2 Solar-Investment Analysis   17
    Three Investment Perspectives, 18; Coping with the
    Uncertain Environment, 24; Dealing with Risk: The Example
    of Domestic Hot-Water Systems, 29; Solar Energy as
    Insurance, 36; Conclusion, 38; Notes, 39

3 Bringing Down the Costs   41
    Progress in Existing Technologies, 41; The Inducement of
    Energy Innovations, 44; Prospects for Solar-Energy Systems,

viii *Solar and Wind Energy*

46; Technological Forecasting, 47; Microforecasts of Technologies on the Drawing Boards, 53; The Behavior of Potential Manufacturers, 57; Notes, 63

4   The Emerging Photovoltaic Markets   64
    Agricultural Photovoltaics, 65; Residential Photovoltaics, 71; The Paradox of Technological Change, 79; Demography and Solar Markets, 82; Conclusions, 82; Notes, 83

5   Farming the Wind   84
    The Supply of Wind Energy, 85; The Economics of the Wind Farm, 88; How Valuable Are the Fuel Savings?, 90; Are There Any Savings in Conventional Capacity?, 101; The Impact on the Volatility of Hourly Loads, 106; Conclusions, 108; Notes, 109

6   Central Power and the Impact of Photovoltaics on Electric Utilities   111
    Centralization Versus Decentralization, 112; How Utilities Select Their Capacity, 114; Capacity Savings: The Impact on the Annual Load Profile, 118; Capacity Savings: The Impact on Reliability, 123; The Impact on the Volatility of Hourly Loads, 125; The Value of Fuel Savings, 127; Renewable Energy as Insurance, 133; Photovoltaics and the Transition to Coal, 134; Conclusions, 137; Notes, 139

7   Conclusions   140
    What Is the Value of Solar Energy?, 143; The Effect of Tax Incentives, 146; Prospects for Technological Innovation, 147; Integrating Solar Energy with Utilities, 148; Solar Energy as Social Insurance, 149; Solar Energy in a Transition to a Postpetroleum Economy, 149; Policy Recommendations, 151

Appendix: Parameters for Cost-Benefit Analysis   154
    Investment Criteria, 155; Overcoming Financing Obstacles: Creative Financing, 157; How to Treat Inflation, 159; What Is the Proper Discount Rate?, 160; How Fast Will Energy Prices Escalate?, 165; BASIC Computer Programs, 168

Bibliography   177
Index   185
Solar Economics software description   189

Figures

3.1 The learning curve for silicon transistors. 44
3.2 Innovation possibilities frontier and achieved isoquants. 45
3.3 The learning curve for photovoltaics. 49
3.4 Economies of scale and market demand. 59
4.1 Discounted benefits and costs per kilowatt, Arizona agriculture. 69
4.2 Discounted benefits and costs per kilowatt, Nebraska agriculture. 70
4.3 Net present worth and present worth discounted back to 1980: Arizona agriculture, pessimistic scenario. 73
4.4 Net present worth of best residential solar systems, 1986: New York City, homeowner's perspective. 78
4.5 Mean daily solar radiation in the United States, kwh/m². 81
5.1 Performance curve: 50-kilowatt windmill system. 87
5.2 Average daily profiles, Fort Worth: Loads and windmill output. 92
5.3 Average daily profiles, Boston: Loads and windmill output. 93
6.1 Costs and the load duration curve. 116
6.2 Average daily load profiles and array output: El Paso. 118
6.3 Average daily load profiles and array output: Boston. 119
6.4 Load duration curves, various levels of solar penetration: Boston Edison. 121
6.5 Effects of photovoltaic penetration on desired mix of generating capacity: Boston Edison. 122

6.6 Weekly peak load and maintenance schedule: Boston Edison. 124
6.7 Loss-of-load probabilities: Effects of conventional and photovoltaic capacity, El Paso and Boston. 126

Tables

2.1	Payback period of solar domestic hot-water system.	33
2.2	Net present value of savings of solar domestic hot-water system.	34
2.3	Net present worth of solar savings, no-solar energy costs, energy costs with solar: Scenario distributions.	37
3.1	Payoff matrix for producer: Industrial strategy vs. size of market.	60
3.2	Payoff matrix for consumer: Cost expectations vs. timing of investment.	61
3.3	Payoff matrix for producer: Timing of entry vs. assessment of rival's technology.	62
4.1	Years in future when rising fuel costs make irrigation unviable.	67
4.2	Irrigation demand and insolation.	68
4.3	Years of demise of conventional irrigation, initial viability, and optimality of photovoltaic irrigation.	72
4.4	Initial year in which residential system becomes economically viable, public-policy perspective.	76
4.5	The effect of tax credits on net present worth of the optimal residential system in New York City.	80
4.6	Population in American insolation belts, 1970 and 1980.	82
5.1	Annual output, capacity factor, and levelized cost of windmill-generated electricity.	89

xii *Solar and Wind Energy*

5.2	Kilowatt-hours of electric generation annually displaced per kilowatt of windmill capacity.	96
5.3	Levelized cost and savings of windmill-generated electricity.	97
5.4	Present worth of fuel savings per kilowatt of windmill capacity: Alternative scenarios, three sites.	99
5.5	Levelized social cost and savings from wind-generated electricity: 3 percent fuel escalation/8 percent discount rate.	101
5.6	Peak and average loads (MW) and load factor at different levels of windmill penetration	102
5.7	Kilowatts of WECS capacity required to displace 1 kilowatt of conventional capacity.	105
5.8	Fluctuations in utility loads under various wind penetrations: Three sites.	107
6.1	Parameters for economic impact analysis.	115
6.2	Peak load, average load, and load factor at different levels of photovoltaic penetration: Five sites.	120
6.3	Peak-load reduction, capacity reduction, load factor: Various penetrations of photovoltaics.	127
6.4	Fluctuations in uti..ty loads under various photovoltaic penetrations: Three sites.	128
6.5	Kilowatt-hours of electric generation displaced per kilowatt of PV capacity.	129
6.6	Levelized cost and savings of PV-generated electricity.	130
6.7	Present worth of fuel savings per kilowatt of photovoltaic capacity: Alternative scenarios, 1980–90, El Paso and Boston.	131
6.8	Levelized costs and benefits of photovoltaics: Four investment perspectives, 1990.	133
6.9	Present worth of savings of photovoltaic capacity: Alternative scenarios, synthetic utilities.	136

Preface

A stabilization of world oil prices has lulled many Americans into believing that the energy problem has evaporated. No perception could be more harmful to our long-term interests. The underlying realities have not changed: The quadrupling of real energy prices since 1973 has stymied economic growth in the West and in nonoil-exporting developing countries; a majority of the world's dwindling oil reserves are concentrated in a few Middle Eastern countries that are, in varying degrees, unstable or hostile to Western interests; and the more abundant sources of depletable energy bear grievous environmental risks. The short-term calm in energy markets should not reduce our vigilance in seeking alternative energy sources that offer an escape from the long-run price escalator, security from the whims of international cartels, and insurance against environmental risks.

Solar energy is often proposed as the panacea, but will it always be the solution of the distant future? I think not. Some technologies for exploiting solar energy, such as passive design, windmills, and domestic hot-water systems, are economically viable now in certain locations. Other technologies, particularly photovoltaics, are likely to enjoy broad-based viability further down the line. How can we soberly assess the near- and long-term prospects for solar energy?

Several types of literature exist on solar energy. There are studies of the technical feasibility of various forms of solar energy, ranging from scholarly discussions of applied physics to passionate excursions into the possibilities of a society based solely on renewable energy resources. There are descriptions of how to build small-scale wind, solar hot-water, or passive heating systems. Little of this literature devotes much attention to whether or not the systems are economically worthwhile. The literature that does deal with the economics of solar energy rarely examines its underlying assumptions, is generally idiosyncratic in terminology, and, more often than not, misleading or incorrect.

This book fills several major gaps in the literature: identifying when and where several solar technologies can profitably penetrate particular market niches, explaining how energy innovations come to be developed and adopted, and examining how weather-dependent technologies can be adopted by a society accustomed to energy available on demand. Understanding these issues is essential to the formulation of an appropriate policy toward the solar-energy option.

To separate polemics from analysis, let me state my stance as a well-wisher to the solar enterprise. I believe that some public expenditure for basic research on solar-energy materials is justifiable, but that private enterprise is better suited to product development, marketing, and retailing than governmental agencies. Publicly supported testing, demonstration, and education in solar-energy systems are essential in lending credibility to the solar industry.

Like most economists, I favor making the social costs of all energy forms apparent to consumers. These costs include those of replacement, environmental protection, and national security. The deregulation of oil and gas is an important step in bringing the prices of these fuels up to replacement cost. Surcharges should be imposed on the extraction, transportation, combustion, and disposal of fuels in proportion to the environmental costs these activities impose upon the public. Tariffs on imported oil would be a vivid signal to consumers of the cost of dependency upon unreliable overseas sources of supply. It is both efficient and equitable that the social costs be passed on to the consumer, not hidden from view. While the deregulation of oil and gas prices is viewed as "pro-oil" and "anticonsumer," nothing will be so favorable to solar technologies (and conservation) as the rise of conventional fuel and electricity prices to their real social costs. In the absence of these "optimal" policies, tax subsidies for renewable energy sources are justified.

My posture as a well-wisher separates me from the advocate who would rush headlong into solar-energy systems without regard to other values, such as monetary cost, human effort or convenience, or from modern Luddites who view solar energy as a way of destroying capitalist industrialism. While oil companies cannot own the sun, they may very well manufacture the best photovoltaic cells that enable us to use the sun's energy. Several defense contractors may develop large-scale windmill systems as a spinoff from their aerospace technology. Whether the breakthroughs are provided by a Japanese conglomerate, a Seven Sister oil company, an aerospace giant, or a fledgling high-technology firm are immaterial to the broad benefits that economical solar-energy technologies could offer humanity.

As a well-wisher of solar energy, I realize that wishing makes it so only in Disney World, not in the real world. To solve our energy problems, solar will have to pass many technical, economic, political, and institutional hurdles. Considerable progress has been made, and thus I am an optimist.

There is a danger of rapid obsolescence in any book that tries to reveal the latest "technological breakthrough of the week." Instead, this book provides a consistent method of analysis of key investment decisions: Would a homeowner find it profitable to purchase a particular solar system in a particular location now? Does it pay to wait until conventional energy costs increase or solar systems become cheaper? When should a manufacturer make a commitment to mass-production facilities that would reduce the cost of solar systems? Can the costs of solar-energy systems be forecast with confidence? When will solar-energy technologies become profitable in a particular market? What are the conventional "avoided costs" a utility can enjoy as a result of solar penetration?

The answers are addressed to the academics, policy makers, and businessmen such as utility executives who are concerned with new energy technologies. Most of the economic analysis is presented with simple arithmetic. The relevant algebraic equations, theoretical discussions, and microcomputer programs are found in the Appendix.

Chapter 1 defines the dimensions of the energy problem and describes the menu of solar-energy technologies that may offer a solution. Because of the great variety of technologies, it is impossible to examine all in great detail. This book focuses on solar hot-water heating and two solar electric technologies of great promise—wind-energy conversion systems and photovoltaics.

Investments in solar-energy systems are evaluated quite differently by a homeowner, a family farmer, an electric utility, a nonutility corporation, a high-income investor seeking a tax shelter, and a public-policy analyst. Businessmen evaluate innovations according to their contributions to the income statement and balance sheet of the firm. This evaluation does not take into account costs and benefits that do not appear on the private income sheet, such as environmental costs or national-security risks. If innovations that appear profitable from a public or social viewpoint do not appear profitable from a private viewpoint, then the public-policy maker may try to influence private decisions by manipulating prices, costs, and taxes or by imposing regulations. To determine which energy innovations should be adopted requires the analysis of investments from both the public and private points of view.

Chapter 2 examines these different investment perspectives and discusses the various risks of solar-energy investments and methods of coping with these risks. The evaluation framework is applied to solar domestic hot-water systems. Solar thermal and photovoltaic technologies are profitable in only a few locations today, but they may become more so as conventional fuel prices rise and as industry experiences technological improvements through progress down the learning curve and economies of scale. The prospects for dramatic cost decreases seem greatest for systems like photovoltaics, which currently cost five to ten times more than prospective energy savings in most markets.

Chapter 3 examines the technological prospects for photovoltaics in detail by breaking the manufacturing process into several stages, each with its own prospects for cost reduction. Mere technical capability is insufficient to induce industry to develop an innovation. The direction of inventive activity is influenced by profit opportunities. Anticipation of specific factor scarcities tends to induce governments to undertake basic research and firms to engage in applied research and development to reduce the demand or to increase the supply of that factor. The prospect of rising oil and gas prices can be expected to induce the expenditure of funds on increasing the supply of more abundant forms of depletable and renewable energy. The magnitude of expenditures is proportional to the likelihood of technical success, which depends upon the state of science and engineering and the size of the potential market. How does one assess this market?

The demand for technological innovations increases with the net benefits of the new technology over the old. Chapter 4 calculates the prospective profitability of photovoltaic systems in agricultural and residential applications at several sites. As system costs decline and conventional fuel costs continue to rise, new markets emerge.

Both Chapters 3 and 4 note that the prospects for rapid technological change presented by photovoltaics may be self-defeating because of strategic decisions made by industry and consumers. Manufacturers face risks that potential customers will hold back in anticipation of an even better product in the near future and that competitors will hold back with superior technologies. Several resolutions to the paradox of self-defeating technological change are suggested.

Many advocates view solar technologies as a means of freeing individuals from dependence upon a complex, interdependent economic system. The prospects for the diffusion of windmills and photovoltaics, however, are brightened when viewed in a utility-interactive mode. Because electrical-energy storage is so expensive, the utility is valuable as a source of backup and as a sink for surplus. How utilities deal with solar cogenerators—what they charge to provide backup and what they offer for surplus electricity—will be a major influence on the profitability of decentrally owned solar systems. This treatment will depend upon the utility's and the public utility commission's perception of the impact of solar energy on the utility.

Within the next few years, several utilities will begin purchasing electricity generated by "wind farms." Chapter 5 assesses the conventional fuel and capacity costs an electric utility can avoid by linking up wind farms at several sites. Profitability is calculated from the viewpoints of the owner of the wind farm, the utility, and the public-policy maker.

Chapter 6 explores the consequences of photovoltaic penetration on the electric utility system. The advantages and disadvantages of centralized versus decentralized solar-energy systems are compared. Avoided fuel and capacity costs are estimated in five sites. The impact of the transition of electric utilities to more coal-fired generation is discussed.

In the concluding chapter, the solar prospect is reviewed and policies for accelerating the development of the solar options are explored.

Acknowledgments

This monograph had its beginnings in 1976 on a pile of biomass. Neighbor Ron Matlin and I were spending a sunny spring afternoon gathering organic matter for our gardens at the Thyme Farm Stables in Lexington, Massachusetts. Between shovelsful, talk turned to different farming practices, and a few idle references were made to the slash-and-burn agriculture I had observed in Brazil. A few months later, Ron was wrapping up his work on satellites at M.I.T. Lincoln Laboratory and turning to a new project that was a spinoff from the space program. The instrumentation on board satellites was powered by cells that generated electricity from sunlight.

As energy issues became more salient, the laboratory became interested in terrestrial applications of these solar cells, or photovoltaics. Ron became assistant director of a group attempting to demonstrate the use of these cells in operating an irrigation pump. While his main mission was to examine engineering feasibility, Ron performed a few obligatory economic calculations on the back of an envelope. Wanting to check them with an expert in agricultural economics, he recalled our barnyard conversations and asked me for a few comments. From simply offering friendly advice, I was eventually drawn more deeply into several of the laboratory's studies. Working with a team of engineers, I undertook economic analyses of photovoltaic applications in broadcasting, residences, and in electric utilities. Ron and I published our early results on agriculture and were astounded by the magnitude of reprint requests from around the world. Agricultural researchers appeared to view photovoltaics as optimistically as we. I was by now addicted to the study of the economics of new energy technologies.

Since I moved to the University of Texas at Dallas, I have learned much about solar energy from colleagues at the School of Natural Sciences, where I now teach, especially from Erven Fenyves and Polykarp Kusch. Kusch made valuable comments on the first three chapters. Erven was gracious enough to review the entire manuscript. Both have been patient in correcting my ignorance of physical laws.

As the ideas in this book incubated, I have benefited greatly from interaction with friends from M.I.T. Lincoln Laboratory, particularly Marvin Pope, Edward Kern, Jr., and Miles Russell. D. Richard Neill and George Curtis of the Hawaii Natural Energy Institute introduced me to wind-energy conversion systems and provided me with invaluable data. I am grateful to Dick Neill and to Frank D. Eldridge for reviewing the chapter on wind farms. Cass Moret made valuable comments on several chapters. Richard D. Tabors and Edward Kern, Jr., helped in refocusing the entire manuscript.

In reviewing two drafts of my manuscript, Jules Levine of Texas Instruments raised my sensitivity to the problems facing potential manufacturers of new energy technologies. Without the computer-science skills of my wife, Arlene, and son, Douglas, the empirical analyses would have been far more difficult. I thank my father, Ira Katzman, for designing the graphs.

I owe a considerable debt to the John Simon Guggenheim Memorial Foundation for enabling me to devote a year to writing the manuscript.

CONVERSION FACTORS

A. Energy

	Calorie	kwh	kJ	Btu
Calorie	1.0	1.163×10^{-6}	4.184×10^{-3}	3.97×10^{-3}
kwh	8.60×10^{5}	1.0	3.6×10^{3}	3410
kJ	239	2.78×10^{-4}	1.0	9.48×10^{-1}
Btu	252	2.93×10^{-4}	1.055	1.0

B. Power Density

	Cal/cm^2	kw/m^2	kJ/m^2	Btu/ft^2
Cal/cm^2	1.0	.01163	41.84	3.69
kw/m^2	85.98	1.0	3.6×10^{3}	3.67×10^{4}
kJ/m^2	.0239	2.78×10^{-4}	1.0	.088
Btu/ft^2	.271	3.155×10^{-3}	11.35	1.0

C. Energy Contents of Fuels

1 bbl. (42 gal.) crude oil	5.8 MBtu (millions of British thermal units)
1 bbl. residual fuel oil	6.3 MBtu
1000 cu. ft. natural gas	1.02-1.09 MBtu
1 ton lignite coal	12-15 MBtu
1 ton anthracite coal	27-30 MBtu

ABBREVIATIONS

BW   *Business Week*
DTH   *Dallas Times-Herald*
EEI   Edison Electric Institute
ELP   *Electric Light and Power*
EPRIJ   *Electric Power Research Institute Journal*
HUD   U.S. Dept. of Housing and Urban Development
MER   *Monthly Energy Review*
NRC   National Research Council
NERC   North American (formerly National) Electric Reliability Council
OECD   Organization for Economic Cooperation and Development
PI   *Photovoltaics International*
SA   *Solar Age*
SE   *Solar Engineering*
SEC   *Solar Engineering and Contracting* (successor to SE)

1
Renewable Energy Sources and the Future

The Dimensions of the Energy Problem

The unleashing of productive forces since the Industrial Revolution depended critically upon the increasing application of energy to machinery. At its initial stages, Western industrialization was based upon renewable energy sources. Wood was practically the sole source of thermal energy in early 19th-century America, and it was a major source until 1890 (EEI, 1976, Table G.1). Wood was supplemented with wind and water, also renewable energy sources, as motive forces for industrial mills (Wade, 1974). The first large-scale factories dotted the line of waterfalls near America's northeastern coast. Since good waterfall sites were limited, factories and workshops increasingly turned to charcoal to power their steam engines, particularly in the Midwestern plains, where waterfalls were rare. Charcoal served as the mainstay of American industry until well after the Civil War (Temin, 1966, chap. 1). Rural America commonly used windmills to pump water, and even to power electric appliances, until the 1930s (Inglis, 1978, chap. 1).

The material abundance of the early 19th-century economy was far greater than that of preindustrial societies. Renewable energy technologies, however, were severely constrained in their capacity to sustain economic expansion. Not only were renewable resources limited in supply, but the costs of equipment to convert wood and wind to forms of energy useful to human beings was expensive.

The availability of cheap fossil fuels had effectively eliminated renewable energy as the workhorse of economically advanced societies by the end of the 19th century. With greater energy content per weight and per volume than wood, coal had across-the-board superiority: It was more transportable and easier to store, and it contained trace elements useful in metallurgical processes. Analogously, oil and gas, which began to assume a visible role in the world's energy supply by the end of 19th century, enjoyed across-the-board superiority to coal. Requiring less expensive equipment, less storage space, and less labor for its utilization, oil and gas had driven coal out of the industrial, commercial, and residential markets by the middle of the 20th century. The tenfold expansion of economic output per capita in the industrialized countries over the past century and the increased consumption of coal, oil, and gas went hand in hand (Kuznets, 1966, Table 2–1).

In the decade since the oil embargo by the Arab OPEC nations, the sustainability of an economy based on the consumption of cheap depletable fuels has been called into question. The public has been bombarded by conflicting diagnoses of the problem and quick panaceas. From this welter of information emerges the consensus that the energy problem has at least three dimensions: cost, national security, and the degradation of the environment.

The post-1973 period stands in stark contrast to the previous fifty years, when the prices of fuels and electricity fell dramatically (Schurr, 1979, Table 3-2; EEI, 1980, Tables 60S and 61S). In terms of purchasing power, oil prices in 1982 are four times higher than in 1972.

At current rates of production, America's recoverable oil and gas resources are likely to be exhausted in thirty or forty years. The world's recoverable oil resources may last for a hundred years if consumption remains stable, but if consumption grows at even 2 percent a year, these resources would be exhausted in about fifty years.[1] Impending exhaustion will surely encourage an intensification of the effort to extract these fuels from deeper and less accessible sites, explore new resources, and research improved extraction technologies. Despite these responses on the supply side, the prices of natural oil and gas are likely to continue rising until they are exhausted. As elaborated in the Appendix, occasional gluts and price decreases will represent only random deviations from this long-run tendency.

Rising fuel costs erode the American standard of living, worsen our balance of payments, and dampen productivity growth. At the mundane level, we now pay more for gasoline and heating oil and for petroleum derivatives like plastics and garden fertilizers than in the 1960s. At the national level, the economy has had to devote increasing amounts of scarce capital and labor to extract a given unit of energy from the ground and to sacrifice more goods in exchange for energy imports.

Dependency upon oil imports raises issues of national security and international power. Higher oil prices have redistributed power to many nations that are not inherently friendly to the West and to some that are openly hostile. The 1973 embargo and its aftermath provide a grim hint of the economic costs of an interruption in oil supply of the advanced nations.

In the early 1980s world oil demand has slackened, and prices have fallen from a high of $37/bbl. in 1981 to about $30/bbl. in 1983. The much vaunted end of OPEC's market power may be a mirage (*BW*, 22 March 1982). OPEC's falling share of world production is most likely a short-term phenomenon because its members possess two-thirds of the world's proved crude oil reserves, more than half of which lie in the Persian Gulf region (Exxon, 1982). On the positive side, many of the OPEC nations have embarked on development plans that are premised on high levels of oil exports. The wake from the Iranian revolution, however, suggests that it would be foolhardy to minimize either the likelihood or the consequences of embargoes in the future, be they intentional actions of a cartel, a byproduct or internal strife in the oil-exporting countries, or a result of a military strike at oil supply lines by an adversary.

In the long run, Western vulnerability may diminish by conservation, the establishment of strategic oil reserves, diversification of sources of oil, and substitution of other fuels for oil. Indeed, these measures are being implemented (BW, 18 July 1983). The completion of cost-effective conservation and fuel substitution takes time, and by the year 2000 Western economies may still depend upon oil for one-third of their energy (versus 50 percent in 1980). About half of the oil is still likely to be supplied by OPEC nations (OECD, 1982).

In the interim, diversification may only replace one form of vulnerability with another. The pipeline project that will deliver natural gas from the energy-rich Soviet Union to Western Europe hardly contributes to the security of the West. Justified or not, fears of nuclear weapons proliferation and internal terrorism have halted or slowed the deployment of the breeder reactor in much of the Western world (Nelkin and Pollak, 1980).

The extraction, transportation, and utilization of depletable energy resources and the disposal of their wastes place a burden on the environment and on a human health. The mining of bulky energy resources, such as uranium, coal, and shale, threatens to scar permanently large land areas and to pollute downstream or underground water supplies. Risks of environmental damage from mining, so common in the past, have been reduced by regulations that encourage careful mining and reclamation practices. The costs of complying with existing regulations have been relatively small, perhaps adding only about 10 percent to the delivered cost of coal (Wilson, 1980).

The transportation and storage of fossil fuels pose additional risks to the public. Deaths and injuries from collisions between autos or school buses and coal trains, explosions in liquid natural gas depots, and oil-besmirched beaches are potential consequences of these exposures.

Combustion appears to pose the most serious health and environmental risks of the fossil-fuel cycle. Combustion releases into the atmosphere particulates, sulfur and nitrogen oxides, carbon monoxide, and other metallic traces, some radioactive. Natural gas is the most benign in this respect; coal, the least. The emissions of air pollutants such as sulfur dioxide and particulates are regulated in most industrialized countries. Several technologies for reducing these emissions are available, including the precleaning of coal or the filtering and scrubbing of the flue gas. Despite the standards of the Clean Air Act, emissions from coal-fired power plants may be responsible for as many as 3,000 fatalities annually and hundreds of thousands of cases of respiratory, cardiac, and pulmonary disease (Ramsay, 1979, chap. 2). In addition, emissions from coal-fired generation may be responsible for as much as $1 billion in local property and crop damage. The accumulation of these oxides acidifies precipitation thousands of miles downwind, further damaging lacustrine ecosystems and reducing forest productivity. The slow but steady substitution of coal for oil and gas will surely increase the exposure of humanity and the environment to these risks.

Some of the environmental costs of combustion are passed on to future generations. There are no practicable technologies for reducing carbon dioxide emissions from the combustion process. The combustion of fossil fuels appears to raise significantly the carbon dioxide concentration in the atmosphere. Climatic modification is a complex, poorly understood process. It is becoming plausible, however, that increasing levels of atmospheric carbon dioxide may cause the uneven warming of the globe, with a consequent desertification of such inland semihumid zones as the American grain belt, the melting of polar ice caps, and the inundation of coastal areas. Thus we may reach the limits of the biosphere's capacity to absorb man's output of carbon dioxide and other effluents from combustion even before the fossil fuels run out. Such limits may even lie within the lifetime of our grandchildren (Hansen et al., 1981).

Air pollution problems have an ominous international facet. While the United States may be secure in the exploitation of its own coal reserves, how will the international community deal with accumulating air pollution resulting from the exploitation of coal resources in China, India, and other poor countries? The inability of the United States to resolve the relatively minor problem of acid rain with friendly Canada portends grave international conflicts over the consequences of coal combustion during the next century.

Conservation, the reduction in energy demand, provides the most environmentally benign, secure, and quickest remedy. In high-income economies, the technological possibilities for conservation are substantial. While many simple actions can save fuel, such as wearing a sweater in the winter, conservation may have costs in terms of other values. An economy that has invested in manufacturing equipment and consumer durables like houses and cars on the presumption of cheap energy has to retool. Retooling takes time and absorbs capital that could be invested elsewhere. Since 1973, American industry has conserved fuel by investing in capital equipment that is energy-saving rather than labor-saving. Historically, Western industrialization was based on energy-using and labor-saving innovations. The reversal of this historic trend since 1973 has undoubtedly contributed to the stagnation of worker productivity (Berndt and Wood, 1979; Jorgenson, 1981). For the poor majority of the population, conservation is a condition of life, not a goal to be achieved. Raising the standards of material comfort for the world's poor requires an increasing supply of energy, not a reduction in demand.

The reason the energy problem appears to be so complex is that there is no fix that is cheap, secure, and environmentally safe here and now. There is no imminent return to the golden age of the 1950s and 1960s, when gasoline and electricity were getting cheaper, when the Western nations controlled the oil supplies, and when the environmental consequences of energy use appeared to be minimal. Every alleged panacea falls short on one dimension or another. The world of the 1980s and beyond faces a burgeoning demand for energy, greater environmental dangers from energy use, and a willful cartel of oil producers that, despite its deep internal divisions and lack of power to sanction its members, has been successful in restricting supply. Two approaches, which are not mutually exclusive, are available: exploiting more fully our depletable fossil-fuel resources, or developing new technologies to exploit renewable energy resources, particularly solar energy.

Solar Energy as Manna

Around solar energy has grown the myth of manna: It is abundant, it is clean, and it is democratic, or amenable to decentralization. It cannot be curtailed by a foreign power. It is seen as a "soft" energy path, easily trod by the average person, using common technologies as a prop (Lovins, 1977). To the American people and many Europeans, it is clearly the Cinderella option (SE, December 1980; Haefele, 1981, pp. 355–60). Less-developed tropical countries, so battered by OPEC price increases, are well endowed with sunlight.

The influx of energy from the sun is enormous. At any moment about 1,400 watts of radiant energy strikes each square meter of the sunward face of the earth's outer atmosphere. About 35 percent of this radiation is reflected back into space. Of the remaining 65 percent, about 190 watts per square meter are absorbed by the atmosphere, and 735 watts per square meter strike the surface (termed *insolation*). About one-tenth of the absorbed energy gives rise to the planet's wind currents (Krenz, 1976, pp. 213-15). The magnitude of insolation dwarfs mankind's energy consumption. The energy inherent in this insolation is decidedly more abundant and more ubiquitous than other sources of renewable energy, like geothermal or tidal.

Annual insolation on the United States (54 million quadrillion Btu or quads) is more than five hundred thousand times greater than annual consumption of about 75 quads. A concrete example illustrates its immensity. A small house in the American Midwest with a south-facing roof area of 1,000 square feet (100 square meters) receives the equivalent of 400 kilowatt-hours (kwh) of solar energy each day, or about 12,000 kwh each month. This contrasts with an average monthly electricity use of fewer than 1,000 kwh (EEI, 1979). If this solar influx could be captured and stored with only 10 percent efficiency, residential electric demands could be met without having to capture any insolation that strikes the surrounding yard. This efficiency is well within reach today.

The Menu of Technologies

There are a number of technologies for exploiting this diffuse flow of solar energy in its direct and indirect manifestations. Traditional technologies include the burning of wood and the prudent siting of dwellings, venerable mechanical devices like the windmill, used by the Persians in 250 B.C., and thermal storage tanks, which date from Roman times. Newer technologies include space-age devices for capturing energy from sunlight, drawing-board conceptions of capturing energy from the ocean's thermal gradients, and experimental mimicking of photosynthesis in the laboratory.

Water Power

The world's hydroelectric potential has been fairly well mapped. In North America, Western Europe, and Japan, a large share of physically available potential has already been exploited. Rising costs of conventional fuels are making hydroelectric sites on smaller, overlooked streams more attractive, and North America's electric utilities are planning a 9 percent expansion of hydroelectric capacity, including "minihydroelectric" dams, in the 1980s. Nevertheless, the hydroelectric share of total generating capacity is expected to fall (NERC, 1983). In tropical countries, only a small proportion of the enormous hydroelectric potential has been exploited. Hydroelectric power will be extremely important in certain regions, but it is not a panacea. Even if complete exploitation were economically efficient, hydroelectricity could meet only a small proportion of current world energy demand, perhaps 15 to 20 percent (Haefele, 1981, Table 6.10). Furthermore, hydroelectric power is not free of catastrophic risk, such as the bursting of dams (Hoyle and Hoyle, 1980, chap. 8). In the United States, environmentalists have frequently opposed the siting of dams on wild rivers or pumped storage facilities in scenic valleys.

Biomass Conversion

Today, populations in Asia and Africa obtain 50 to 90 percent of their energy from biomass. In the United States, approximately 10 percent of households utilize wood for a portion of their heating load (L. Brown, 1981, p. 205). Despite these impressive numbers, biomass offers little immediate promise for expanding energy supplies in poor countries or replacing depletable energy in most advanced countries.

The major limitation of biomass as an energy source is its lack of physical abundance. Most plants are able to convert less than 1 percent of incident insolation into biomass averaged over the year. A few exceptional plants, such as sugar cane, attain conversion efficiencies of as much as 3.3 percent during their periods of most rapid growth (Calvin, 1974). In advanced countries, the growth of biomass is simply insufficient to keep up with energy demand. Energy crises occasioned by wood shortages plagued 16th-century England and 18th-century France and Sweden (Berg, 1978). The gross energy content of annual plant growth in the United States is only 25 quads, about one-third of current energy use. If all of the growth were devoted to energy production rather than to food, fiber, or building materials, the net energy yield of biomass would be far less than that figure because of energy expenditures in harvesting. The conversion of biomass to heat by direct combustion, or to liquid and gaseous fuels, would reduce the net yield even further.

A biomass energy base is not without environmental hazards. For example, the burning of wood in dense settlements can cause severe localized air pollution. The crackling fire and aroma of burning spruce that epitomizes the cozy home in the woods of Maine or Sweden signifies ecological disaster in Nepal. Harvesting wood in excess of sustainable yield has led to deforestation, resulting in erosion, desertification, and flooding. For most of the world, the traditional biomass economy, based on a hunter-gatherer technology, is simply unsustainable.

While extraction of energy from current forms of biomass is far from a panacea, biomass can continue to make a modest contribution to the world energy picture and a significant contribution in certain regions and end-use sectors.

First, some residues that are normally collected and then discarded can be more fully exploited. For example, the gross energy content of municipal wastes and organic effluents from American agroindustries like slaughterhouses or feedlots is about 2 quads (Burwell, 1978). Utilities that have attempted to generate electricity from municipal garbage have met failure more often than success (Smock, 1981 b). Remote lumber and paper mills commonly burn waste chips in their boilers, and some sugar mills burn bagasse. Burlington, Vermont, is planning a 50-MW power plant that will be fired by wood wastes (ELP, March 1982). Utilities in Florida and Hawaii are planning to purchase a small amount of electricity generated from sugar-mill bagasse. Such burning of biomass currently generates 1 quad, about half as much energy as America's nuclear plants (Metz and Hammond, 1978, p. xii). This niche seems secure and may even expand, but it is unlikely to contribute much in advanced industrial countries. Furthermore, generating electricity with these wastes is not cheap. Burlington Electric's plant is expected to cost $1,600/kw. A similar plant in California will cost $2,000/kw (SE, November 1980, p. 46), about twice as much as a coal-fired generator.

Second, uncollected crop residues and animal wastes can be converted to usable energy by direct combustion, or by conversion to liquid or gaseous fuel by small-scale digesters. In the United States, residues may yield about 2 to 4 quads net (Burwell, 1978; Pimentel, 1974, 1981). This energy represents only 30 percent of the gross energy content of residues, after subtracting energy expended on collection and processing.

There are serious consequences in removing these residues from the land, for they recycle essential trace elements, prevent erosion by water runoff, and maintain the friability of the soil. These consequences can be offset by returning the remaining solids to the soil, increasing chemical fertilization, altering practices of plowing (no-till sowing), and carefully grading the soil. In poor countries, where the cost of chemical fertilizers, herbicides, and tractors is high, the removal of these wastes may be ecologically disastrous. When the labor and capital costs of collecting residues, refurbishing the soil, and converting the biomass to usable energy are taken into account, the net energy yield may be positive, but the net economic benefits are questionable. Whether residues are best used economically as soil additives or as energy feedstocks is not yet clear. The practice of converting farm residues into gas in China, where labor is relatively cheap, appears to be losing favor (Smil, 1982).

Finally, "energy plantations" could be established on lands marginal to agriculture or offshore, where kelp farming looks promising (North, 1977; Ryther, 1978; Klass, 1978). The plantation crop can then be converted to liquid or gaseous fuel by one of several well-known processes. The most concerted effort at energy farming is in Brazil. The Brazilian program provides incentives for the expansion of sugar-cane plantations and the construction of centralized distilleries. Currently, refiners can derive ethanol from cane sugar for about $33/bbl., of which raw-material costs account for $26/bbl. Because of the complex capital subsidies embedded in the alcohol program, it is difficult to evaluate its true social costs (Goldemberg, 1978, 1981; Rothman et al., 1983). A hidden social cost is the relative increase in the cost of basic foodstuffs, which must compete with sugar cane for prime farmland and capital (L. Brown, 1980).

By no means have we achieved the limits of extracting energy from biomass. Crops have been bred for disease resistance, height, yield, and other desirable characteristics, but little effort has been devoted to maximizing the efficiency of biomass energy capture. An exciting possibility opened up by the revolution in genetic engineering is the improvement of photosynthetic efficiency, perhaps up to the theoretical maximum of about 7 percent year round. Biochemist John Abelson (1980, p. 1321) muses:

> I believe that the most important and pressing applications [of genetic engineering] will come in providing solutions to our energy problems. Photosynthesis is a solution to the need for a renewable energy source that has been perfecting itself for more than 3 billion years. Almost certainly an important component to the solution of the energy problem will have to come from biology. . . . It should be evident that many of the tools for this task are already at hand.

The revolution in genetic engineering has attracted enormous sums of private funding for research and development efforts that are likely to spill over into energy research. Several oil and chemical corporations have acquired seed companies in order to develop and market new agricultural products. It would not be surprising to see these subsidiaries spawning new biological substitutes for fuels and chemical feedstocks.[2] A breakthrough in photosynthetic conversion would totally alter the economics of energy extraction from biomass, but speculation on the economics would be premature.

Passive Heating and Cooling Systems

Passive solar systems include some of mankind's oldest techniques for taking advantage of the sun's warmth when the weather is cold and shading when it is hot (Butti and Perlin, 1980). Such systems are passive in the sense that they use very little specialized mechanical equipment.

Often the line between passive systems and conservation is rather gray. For example, most people in the world dry their clothing in the sun. This old-fashioned passive system uses little capital equipment (perhaps a clothesline) and no fossil fuel, at a cost in time and effort. Given the value of time, few Americans would regard the abandonment of automatic clothes dryers as cost-effective.

Passive solar systems can be incorporated into the buildings by clever, site-specific design. For example, a well-designed residence in the North, with extensive heating demands and minimal cooling demands, would differ in orientation and materials from a dwelling in the Southeast, where cooling and dehumidification are more important. In contemporary northern dwellings, large south-facing windows and wall insulation can capture and retain heat from the winter sun. In the Southeast, the roof overhang must be designed to block the high summer sun from these south-facing windows.

Such passive-solar design practices as proper siting, landscaping, and materials selection can be economical because they cost little more in labor and materials than poor design (Cassiday, 1978; Snell, Achenbach, and Peterson, 1978; OTA, 1980). Unfortunately, homebuilders are generally unaware of passive-design principles, and the acquisition and dissemination of knowledge of these practices is not free.

More ambiguous are the net payoffs from complex passive systems that incorporate additional materials like greenhouses, thermal storage tanks, or massive Trombe walls. Because passive systems are not standardized, mass-produced products, it is hard to generalize about their costs. Heat-absorbing Trombe walls can cost $1,500 and small heat-retaining greenhouses about $2,000 (SA, May 1981). The value of heat savings from Trombe walls and greenhouses in New Mexico are about the same as the cost of these structures (Schurr et al., 1979, p. 325). After examining five sites around the United States, Shapira, Brite and Yost

(1981) found that there are no net savings to building a house partially underground. Unfortunately, the economic assumptions are not reported. A detailed simulation of passive design for a Colorado house found that the cost of window overhangs, window shutters, triple glazing, and increased mass for thermal storage did not pay (Feldman and Wirtshafter, 1980, chap. 4). How representative these results are is not clear.

Obviously, it is difficult to alter the siting and orientation of existing houses, but passive-solar design concepts can be incorporated into newly constructed dwellings. In the year 2000, about one-third of the nation's housing units will have been built since 1980. This percentage will be higher in the rapidly growing Sunbelt. Without a doubt, passive-solar design will fill a major niche in heating and cooling buildings, which currently consume nearly 20 percent of the American energy budget. Oak Ridge National Laboratory projects that the incorporation of cost-effective building design in new housing can reduce total residential space-conditioning demands by about 20 percent by the year 2000, despite a 50 percent increase in the housing stock (Schurr, 1979, pp. 75, 127–48; Hirst and Carney, 1977).

Active Thermal Systems

Active solar-thermal systems utilize large flat plates ribbed by tubes to collect heat. The fluid in the tubes is then transported to a thermal storage device or to an end use. For example, a domestic hot-water system may utilize water or ethylene glycol to transport the heat from the collector to a heat exchanger in a storage tank. The heat from the collector (but not the liquid itself) flows through the walls of the exchanger and heats the water. Such systems can deliver to the storage tank about 40 percent of the energy incident on the collectors.

Requiring more elaborate and specialized hardware, active systems are far more expensive than passive systems, which embody little more than slight design changes in common construction practices. Simple flat-plate collectors can heat fluids to about 175 degrees F/80 degrees C, which is suitable for space-heating systems (Metz and Hammond, 1978, chap. 4). The flat-plate collector is clearly capable of meeting heating requirements for domestic hot water (130 degrees F/54 degrees C) and for swimming pools (80 degrees F/26 degrees C).

Simple reflectors can provide the concentration necessary for boiling water and for other cooking needs. Such systems may meet the needs of rural populations in low-income countries that are unlikely to be served by central utilities in the near future (Daniels, 1964, chap. 5).

Intermediate temperatures of 300 to 500 degrees F/160 to 260 degrees C, adequate for many industrial processes, can be achieved by concentrating mirrors. Theoretically, 45 percent of the energy in insolation can be captured by advanced systems (K. Brown, 1981). Continuous eight-hour or even twenty-four-hour operation of factories requires considerable backup power, for thermal storage at such temperatures is difficult (Rapp, 1981, pp. 15-16).

In theory, concentrators can achieve temperatures of 2,500 degrees F/ 1,370 degrees C (Rapp, 1981, pp. 322-24). Sun-tracking mirrors, or "heliostats," can maintain high temperatures at the focal point even when the sun is low. In practice, a thousandfold concentration, bringing temperatures up to 900 degrees F/500 degrees C, has been achieved in the "power tower." Heliostats covering hundreds of acres focus sunlight on an elevated conventional steam boiler that runs an electric generator. The concept is now being demonstrated at several sites, the largest of which is Southern California Edison's 10-MW Solar One. The plant's solar components alone (heliostats, tower, storage) cost about $12,000/ kw, compared to $800/kw for a coal-fired generating plant. The materials and technology of the power-tower concept are conventional—lots of concrete and steel—and unlikely to realize cost-reducing improvements. Experience in construction and design, however, is expected to reduce costs eventually to about $6,500/kw (ELP, June 1982, p. 58; Power, April 1983, p. 118).

The performance characteristics of solar-thermal systems are influenced by the amount of complementary storage. Heat captured during the daily insolation peak can be stored in massive containers filled with rocks or water for discharge during the night. Larger storage systems could meet low- and moderate-temperature energy needs over several cloudy days. Extremely large underground thermal-storage devices could conceivably store heat from season to season. Chemicals that store heat when they change from liquid to solid are another possibility for thermal storage. Whatever the technical possibilities, storage devices are expensive, and how much storage is desirable becomes an economic decision. Unfortunately, no major cost-reducing breakthroughs in either domestic hot-water systems or thermal storage appear to be on the horizon (Jacobsen and Ackerman, 1981).

Wind Power

About 1,500 quads of wind energy sweep across the United States each year. Most of this energy exists at high altitudes, but about 600 quads brush close to the surface, and an additional 60 quads is available within fifteen miles offshore (Eldridge, 1980, pp. 3–7, 181). If all the potential sites for mills were exploited, wind might supply as much as one-third of the current energy demand in the world (Haefele, 1981, Table 6.10;

Eldridge, 1980, chap. 9). How much of this potential will actually be exploited depends upon windmill economics.

Wind is less "democratic" than other forms of solar energy in that it is less evenly dispersed than sunlight. While the annual energy influx from sunlight varies by less than a factor of two between the least favorable sites in the Northeast and the most favorable sites in the Southwest, annual wind energy varies by a factor of more than five. While average annual insolation varies in broad bands as a function of latitude and humidity or cloud cover, wind energy is heavily influenced by local topography. Wind speed is relatively great on unobstructed hilltops and ridges, in mountain passes, and on flat regions where there is little surface friction, like grassy plains or seacoasts. Sites only a few miles apart may vary by an order of magnitude in available wind power. The areas of peak wind power in the United States include the mountains of Colorado, the plains of Kansas, the Adirondacks, and Cape Cod. Within these regions, the windiest sites may not be the best from an economic point of view. First, they may be far from transmission lines or from load centers. Second, sites where the combined cost of generation and transmission are minimized may be subject to competing recreational or scenic values.

Like insolation, wind speed is not uniformly distributed over time. At most sites, daily and annual cycles of average wind speed are quite stable from year to year. There is considerable variability around the average, though. In particularly windy sites, speeds can change by 50 percent from one hour to the next (Cheng and Wong, 1979). Gusting may alter wind speeds by a factor of three or more in a matter of seconds. Because wind energy varies with the cube of wind speed, the latter fluctuations are amplified in the power grid.

In the past, windmills have been commonly used to pump water and to power gristmills or sawmills. When the wind is strong, water is drawn up to a storage tank to be used when needed. For these simple applications, a close phasing between temporal patterns of wind power and energy use is not essential. Unfortunately, many electricity demands require a closer phasing; in the early evening when lights may be needed there may be no wind. Using windmills for electric generation requires either some form of electric storage or backup by another energy system.

Although rooted in an ancient technology, modern aerodynamic principles and construction materials are currently being applied to more efficient and cheaper windmills. Wind-energy conversion systems currently cost about $2,000 to 4,500/kw, about two to five times more than a coal-fired electric plant. Even without major cost decreases, wind energy seems to be economical in a few sites today. Economies of mass production may result in a substantial market share for wind energy within the decade.

Photovoltaics

The solar technology that offers the greatest long-term promise for meeting nearly the entire gamut of energy needs is photovoltaics, which uses solid-state cells to convert sunlight into the most flexible form of energy—electricity. Sunlight is emitted as packets of energy called photons. The energy content of these photons varies directly with their frequency. Different frequencies in the visible portion of the spectrum are perceived as colors. Photons with sufficient energy can dislodge electrons from the atoms of certain materials, exemplifying the "photovoltaic effect." These free electrons can be collected and channeled through an external wire, thus creating an electric current.

The solids most susceptible to the photovoltaic effect are called semiconductors. While there are many other semiconducting materials, such as gallium arsenide and cadmium sulfide, today silicon is most commonly used in the manufacture of solar cells because of the maturity of silicon technology in the electronics industry, silicon's abundance on the earth's crust, and its responsiveness to the sun's spectrum.

As a semiconductor, the photovoltaic cell is akin to the transistor and microprocessor, the epitome of modern technology. Simple photovoltaic systems include large, thin sheets of specially wired semiconductors and devices for regulating the voltage. These systems generate direct current, as do flashlight or automobile batteries. Complete systems include inverters, electronic devices that convert direct current (DC) into the alternating current (AC) commonly used in household electrical circuits.

An advantage of photovoltaics is their modularity. They are about as cheap to operate in 100-watt units as in 100-kilowatt units. Another advantage is their capacity to function alone, without being connected to the electric grid. Because the extension of electric transmission and distribution lines is so costly, photovoltaics can compete in remote applications. Thus they have found their initial commercial markets in such applications as the cathodic protection of deep oil wells against corrosion, relaying radio signals, and agricultural pumping. Photovoltaics may provide the earliest hope for the electrification of remote villages in developing countries.

Like windmills, photovoltaics pose the problem of reconciling the correlation between energy supply and demand. While windmills may produce power at night, terrestrial photovoltaics will not. They must be backed up by either a utility linkage or electric storage.

Photovoltaic systems can by no means capture 100 percent of the solar energy falling on a given surface. One kilowatt-hour of solar energy striking one square meter of photovoltaic material generates only about one-tenth kilowatt-hour of electricity. More relevant than conversion efficiency is the cost of solar electricity. Originally developed to power space satellites, photovoltaics were designed to meet high performance standards, with minor attention to cost. A photovoltaic array with one kilowatt of generating capacity cost about $100,000 in the early 1960s. Semiconductor devices like the transistor have experienced cost reductions on the order of 25 percent a year for about two decades. The photovoltaic industry has made comparable strides in reducing costs in the 1970s. As yet, the production of photovoltaic panels is analogous to a cottage industry, capable of enjoying major cost decreases as a result of mass production.

In 1980 cells cost about $10,000/kw; the balance of the system, another $5,000/kw. By 1983 the cost of cells had fallen to less than $5,000/kw in 1980 dollars, but little progress had been made in reducing the costs on the balance. The $10,000/kw capital cost of photovoltaic systems exceeds by far that of electric generators based on solar thermal ($6,500/kw), wind ($2,000–4,500/kw), or coal-fired systems ($800/kw). Offsetting these high capital costs of photovoltaics are zero fuel costs, little maintenance, high reliability, minimal environmental impact, and nearly universal applicability at any scale. Moreover, unlike these other technologies, photovoltaics enjoy considerable potential to become significantly less expensive.

Is Solar Energy a Panacea?

Despite the abundance of solar energy, it is wishful thinking to regard it as a panacea. Unlike manna, solar energy cannot be consumed without cost. While the "fuel" is free, the devices for capturing or transforming solar energy into the various forms of energy demanded in modern civilization are not. For the amount of energy they deliver, most of these devices are costly and themselves consume much energy to build. The major challenge in exploiting this vast supply of solar energy is to be able to modernize traditional technologies and develop new technologies that are consistent with a high material standard of living. The prospects for solar technologies are brighter the greater their amenability to cost reduction through the spread of mass production and through gains in manufacturing experience.

Not all solar technologies lend themselves to backyard installation by the weekend handyman. Indeed, exploiting some manifestations of solar energy are simple or traditional, like a wood stove, a waterwheel, or a windmill. But the solar technologies that will add most to the world's energy supply in the near future are small hydroelectric plants and large windmills, whose costs are well beyond the means of the average family. In the longer run, the most significant prospects for expanding the supply of renewable energy appear to lie in the realm of high-technology, heavily industrialized processes, such as photovoltaic cell manufacture.

Finally, solar-energy systems are not completely free of health risks. Catastrophes from systems such as photovoltaics or windmills appear unlikely. Some raw materials for solar cells, like gallium arsenide, are toxic. Professional workmen and amateur handymen are subject to injury in manufacturing, installing, and maintaining solar equipment. If the risks from decentralized energy systems are similar to those from decentralized transportation systems, epitomized by the automobile, these risks are not negligible (Neff, 1981).

The intermittency of sunlight and wind poses no insurmountable obstacle to the adoption of photovoltaics or windmills in a society accustomed to energy upon demand if they are supplemented by conventional power. For these solar technologies to supplant conventional energy totally, the problem of cheap thermal, chemical, or electric energy storage must be solved.

Notes

1. The world's ultimately recoverable oil resources have been estimated at about 9,800 to 12,800 quads and corresponding American resources at about 670 to 800 quads (Schurr et al., 1979, Table 7-1; CONAES, 1980, Table 3-1; Haefele, 1981, p. 57). If American oil production remained steady at about 18 quads, resources would be exhausted in thirty-seven to forty-four years. If world oil production remained steady at 125 quads, the oil would last nearly 100 years. While oil demand in the industrialized world is expected to stabilize, rising demand in less developed countries is expected to drive world demand up 2 to 3 percent per year. At these rates, the world's recoverable oil resources would be exhausted in about fifty years. Because natural gas is difficult to transport overseas, the American recoverable resource base of 700 to 900 quads is relevant to our analysis. If demand remained steady at 20 quads, the resources would serve for about forty-five years, but a 3 percent annual growth in demand would bring Doomsday about twenty years closer.

2. See "Biotechnology Boom Reaches Agriculture," *Science* 213 (18 September 1981): 1339–41.

2
Solar-Investment Analysis

While they may be embodied in new materials and new products, solar-energy technologies are process innovations. Their primary task is to perform conventional functions, such as providing warmth or electricity, in unconventional ways and with minimal harmful side effects. Process innovations are easily analyzed by standard investment principles.

All solar-energy systems entail large, immediate expenditures on equipment for the prospect of saving conventional energy for many years to come. The analysis of solar-energy systems involves two important complications. First, because of the exceptional degree of regulation and tax preferences in the energy field, numerical results are highly dependent upon the tax status of the potential investor. Second, investments in solar-energy technologies are generally subject to greater uncertainties than conventional energy investments. These distinguishing features aside, the general approach to investment analysis is the same for all types of solar-energy systems, ranging from 5-kw residential photovoltaics and 5-MW wind farms to 500-MW hydroelectric dams. The approach is illustrated here by a solar domestic hot-water system and in subsequent chapters by more complex systems. A technical discussion of the algebraic derivations, selection of the key parameters, and computer programs can be found in the Appendix.

Three Investment Perspectives

The arithmetic of investment analysis looks quite different from the perspectives of the public-policy maker, commercial entrepreneur, or the homeowner. The differences lay in the treatment of prices, taxes, depreciation, and discounting. From all three perspectives, the benefits of solar-energy systems are proportional to the cost of the energy saved. Specifying this cost is not trivial. A good benchmark cost should reflect an equilibrium—a fairly stable, market-clearing price.

Does such a price exist in today's volatile markets, buffeted by the twin oil shocks of the 1970s and the glut during the world recession of the early 1980s? The average barrel of oil imported to the United States cost about $34 in 1980, $37 in 1981, $34 in 1982, and then dipped as low as $28 in early 1983 (MER, July 1983). Expressed in 1980 dollars, the prices in the early 1980s average to a little less than $6/MBtu. From this benchmark, all other values are derived. For example:

> Low-sulfur, No. 6 residual fuel oil for utility boilers costs a little more than crude, about $6/MBtu. While utilities use less expensive, high-sulfur fuel oil, the required scrubbing of the flue gas can bring the cost up to $6/MBtu. Small (10 to 20 percent) interregional variations in prices are diminishing rapidly and are ignored.

> The refining and distribution of home-heating oil adds about $3/MBtu to the price of crude. This brings the delivered price to $9/MBtu.

The Public-Policy Perspective

The public-policy perspective is a normative viewpoint in which social benefits and social costs of investments are compared. This perspective generally differs from the actual behavior of federal policy makers who undertake project evaluation, but it is nevertheless a useful standard.[1]

The market prices of solar-energy systems reflect their social costs fairly well. However, there may be additional hidden costs that may not involve any direct cash outlays. For example, if solar arrays were installed on the lawn, they would diminish the space for recreation and gardening. Such hidden costs are likely to be given some value by homeowners and should be included in any calculation.

The social cost of conventional energy reflects the costs of replacement, environmental degradation, and national-security risks. In practice, market prices of fuels have not reflected these costs well. Historically, through tax incentives and the public assumption of liabilities, the federal government has subsidized the production of nuclear fuel, coal, and oil.

Price and quantity restrictions create a gap between prices in the domestic and international market, which can be viewed as the replacement source. Until 1973, quotas on imported oil resulted in American prices exceeding world prices. With the imposition of price controls in the 1973–81 period, domestic oil producers were no longer subsidized but instead were covertly taxed. Marginal supplies were imported at world prices, which were about twice the controlled domestic prices. Price controls on natural gas traded on interstate markets result in delivered gas prices averaging about one-half of the price of newly discovered gas, the marginal supply. The partial deregulation enacted in the 1978 Natural Gas Policy Act maintains controls on "old" natural gas (discovered before 1973) and hence on the discrepancy between price and replacement cost.

An important element in the social cost of fuels is the occupational, health, and environmental costs of mining, transportation, combustion, and disposal. In some cases, damage to employees and third parties (external costs) has been recovered by the tort process. To the extent damage suits are successful, these costs are taken into account by the energy sector. In most cases, the damage is so diffuse that the tort process is ineffective, and regulation has proven necessary.

Of all fossil-fuel cycles, that of coal has historically inflicted the widest range of external costs. The regulation of mine safety, land reclamation, transportation of hazardous materials, and air emissions have been only partially successful in placing these costs on the balance sheet of the energy industry. The Clean Air Act requires industries that burn coal and oil to meet stringent limitations on emissions of sulfur, particulates, and oxides of nitrogen. While the cost of these regulations are "internalized" through outlays on inherently clean fuel, on cleaning the fuel on site, or on scrubbing the flue gas, residual emissions are significant. Because of the large ecological and epidemiological uncertainties and political judgments involved, it is difficult, if not impossible, to pinpoint the environmental costs of coal combustion.

One estimate of the health and property damage from sulfur dioxide emissions alone has been estimated at about 20 cents/MBtu or 2 mills/kwh (Perl and Dunbar, 1982). These calculations ignore the damage to crops and recreational values from acid precipitation and from other pollutants, namely oxides of nitrogen (NRC, 1983). Furthermore, carbon dioxide emissions from power plants, which are not regulated in the United States, are contributing to the potentially dangerous atmospheric "greenhouse effect." While the cost of acid rain has not been computed, proposals to eliminate one-half of the sulfur dioxide emissions would add about one mill to the cost of each kilowatt-hour generated by fossil fuel (ELP, August 1981). The World Coal Study estimates that total elimination of these oxides would add about $1/MBtu to the cost of coal (Wilson, 1980, Table 4-5).

National-security costs do not appear in the price of oil. At a minimum, these costs reflect the probability of an embargo, multiplied by the resulting losses to the economy. Estimates made in the late 1970s of the national-security costs of imported oil vary by a factor of 10, from $4/bbl. to $40/bbl. (Landsberg, 1979, chap. 6; Plummer, 1981; Rowen and Weyant, 1982). These figures suggest imputing a surcharge of 10 to 100 percent on the market price of imported oil. With the buildup of strategic petroleum reserves, fuel substitution, and conservation, the probability of an embargo and the magnitude of the probable disruption appear to have diminished. If so, then the national-security premium may be closer to the lower end of the spectrum.

One cannot overemphasize that the environmental and security consequences of different energy futures are difficult to discern, and individuals will differ in their evaluations of these consequences. Nevertheless, these external costs of conventional energy cannot be ignored in public investment analysis. One should not be preoccupied by the spurious search for five-digit accuracy in their measurement. At best, scientists and technicians can delimit the order of magnitude of these external costs, but in actuality they are determined through the political process because they involve ethical concerns, such as the valuation of human life and the welfare of future generations. This task is not beyond the capability of a democratic political system, which implicitly makes intergenerational-welfare comparisons and judgments about life and death when it writes environmental regulations or licenses a nuclear plant. While public policy since 1977 has moved energy prices closer to real social costs, there is still a considerable distance to traverse.

The social costs used in this analysis are the marginal fuel prices under a scenario of complete natural gas deregulation, gradual tightening of air-quality regulations on coal emissions, and implementation of an acceptable nuclear waste disposal scheme.

As noted, the marginal cost of gas is generally higher than the retail market price, which may reflect a mix of old and new contracts. As a direct competitor of oil in many markets, gas is likely to be similar in price. Its advantage as a cleaner boiler fuel may be offset by its greater difficulty in storing. A marginal cost of $6/MBtu is imputed to natural gas in the producing states of the Southwest. An additional $1.50/MBtu is added for transmission and distribution for consumers in the West and North. The national average market price of natural gas delivered to residences is about $4.50/MBtu and to utilities (mostly in the

Southwest) about $2/MBtu. Much of the interregional difference in gas price can be attributed to regulation rather than to real costs.

In several scenarios, public-policy analysts impute a 25 percent national-security surcharge to residual fuel oil. Because natural gas can displace oil, it earns the same 25 percent national-security premium. This brings the social value of oil and gas displaced up to $7.50/MBtu.

Because of its heterogeneous quality and high transportation costs, coal cannot easily be given a national price. On the average, coal delivered to American utilities cost about $1.50/MBtu in 1980. Interregionally, the price ranged from about $2 in New England, northwestern Europe, and Japan to less than $1 in the Mountain States (MER, July 1983; Wilson, 1980, Table 3-1). These figures provide a minimum estimate of the social costs of this fuel. The environmental costs add a minimum of 20 cents/MBtu (a 10 percent surcharge), but if atmospheric carbon dioxide accumulation is taken into account, policy analysts can make a persuasive case for a higher surcharge, say 25 percent.

The cost of nuclear energy has proven to be extremely difficult to forecast. Construction delays that result from technical and political difficulties have led to unforeseen escalation in plant costs. Because solar and wind systems are unlikely to save much nuclear plant capacity, the cost of nuclear fuel is more relevant. As with the costs of other energy sources, some of the costs of nuclear energy are borne by the public. Under the Price-Anderson Act, the public assumes all risks of nuclear accidents beyond $560 million. Nuclear plant operators have yet to bear the costs of permanent waste disposal.

The long-term cost of the complete nuclear fuel cycle was estimated by Landsberg et al. (1979, p. 419). These costs include .32 cents/kwh for uranium oxide mining, .01 cents/kwh for conversion to uranium fluoride, .20 cents/kwh for enrichment, .05 cents/kwh for fabrication of pellets, and .08 cents/kwh for waste management. These costs total .66 cents/kwh in 1978 dollars, or .8 cents/kwh in 1980 dollars. Rounding off the figures, we arrive at $1/MBtu as the minimum bound of the social costs, excluding public subsidy of catastrophic risks.

The specification of the proper discount rate for social-benefit cost analysis has been subject to considerable controversy (Lind, 1983a, 1983b). The widely accepted view is that the discount rate should reflect aftertax earnings on savings, which is approximately 5 percent above inflation. When using this discount rate, however, costs should be adjusted to reflect the alternative value of investment funds, either in private investment or in consumption. The suggested correction factors are equivalent to a 10 percent discount rate.[2] The 10 percent rate has been mandated for federal projects by the Office of Management and Budget, but a somewhat lower rate is used for federal water projects

(Lind, 1983a). The 5 and 10 percent rates are used here as reasonable bounds.

The Private-Business Perspective

In assessing investments in solar-energy systems, profit-making businesses take into account several elements that are ignored in the public-policy calculus: constraints on prices, investment tax credits, tax deductibility of interest, and accelerated depreciation. By definition, the private business that uses or produces energy ignores the external costs of conventional energy. It considers only market prices, many of which are controlled.

The market price of electricity is currently below replacement cost. This means that the social value of electricity saved by a renewable energy system may exceed the reduction in a customer's utility bill. Regulatory commissions have not permitted utilities to recover the replacement cost of their older plants or to charge rates that reflect costs of acquiring new sources of fuel or generating plants. Rather, all costs, old and new, are rolled together. Consequently, the costs of building new capacity exceeds perhaps by a factor of two the average costs utilities can charge.

By longstanding practice, utilities tend to charge a fixed rate for electric use, regardless of when the use occurs. The utility is likely to burn expensive oil or gas during the peak of the daily load cycle and inexpensive coal during the night. Electric rates generally average the expensive and cheap fuels. A solar-energy system that produced hot water during the daytime peak might be saving the utility 8 cents/kwh, but a customer's electric bill may drop only by 5 cents/kwh.

These discrepancies between the rates that utilities charge and the cost savings realized by reductions in load can be substantial, and they explain why some utilities have begun subsidizing the installation of solar-energy systems or insulation.

Federal tax laws allow a 10 percent tax credit for any tangible investment. The Crude Oil Windfall Profits Tax Act of 1980 allows nonutility businesses an additional 15 percent credit for investments in renewable energy systems. The differential eligibility for the renewable-energy tax credit has created a profitable opportunity for nonutility businesses to sell renewable energy to a utility (see Chapter 5). As an offset to these tax advantages, businesses pay taxes on incremental profits earned as a result of reducing conventional energy bills.

The economic depreciation of a solar-energy system equals the deterioration in performance or obsolescence reflecting the development of superior or cheaper devices. This is quite different from the standard accounting concept of depreciation. In computing taxable income, businesses are permitted to depreciate equipment at a faster rate than true economic depreciation. Accelerated depreciation permits businesses to recover the capital value of their investment early in the life cycle, when the discount factor is small. The Economic Recovery Act of 1981 allows nonutility businesses to depreciate investments in renewable energy systems in five years and utilities in ten years.

If conventional energy were priced at social cost, then the net effects of the tax system on the evaluation of solar investments by businesses would be minimal. As shown in the examples below, taxes reduce benefits by the businesses' income and property tax rate, perhaps up to 50 percent, while tax preferences can reduce costs by about 50 percent for utilities and by even more for nonutility businesses.

Investor-owned utilities, which comprise three-quarters of installed American generating capacity, finance investments by issuing bonds and stocks in nearly equal proportions. The interest paid on bonds by investor-owned utilities is deductible from corporate profit. The tax deductibility of interest effectively lowers the interest rate on bonds (see Appendix).

The interest rate according to which utilities should discount investments is a weighted sum of the aftertax interest on bonds and the return on stock. Subtracting inflation, this is a real rate of about 5 percent (Lind, 1983b; Stauffer, 1983). In fact, nearly half of the utilities use the pretax cost of equity, which is close to 10 percent (Corey, 1983). Again, 5 and 10 percent rates are used as bounds.

To add to the complication, accelerated depreciation, tax credits, and the tax deductibility of interest are irrelevant to municipal utilities, which do not pay federal corporate income taxes. Municipal utilities, however, can issue tax-free revenue bonds, which generally reduce interest charges by about one-fourth. An appropriate discount rate for municipal utilities is close to 5 percent.

The Homeowner's Perspective

Homeowners are eligible to deduct interest payments from their taxable income and enjoy a 40 percent income-tax credit for solar investments, but they are not entitled to any depreciation allowance.

In theory, homeowners should discount energy-saving investments at the aftertax interest rate earned on savings, which is about 5 percent. In actuality, they act as if they use interest rates as high as 15 to 25 percent (Hausman, 1979), which is equivalent to demanding a short payback period. This behavior is hardly irrational. Homeowners move frequently and are uncertain about the resale value of solar equipment.[3] The high

24 *Solar and Wind Energy*

discount rate may also reflect uncertainty about the reliability of the equipment.

Coping with the Uncertain Environment

Any investment is risky, and all investments share common uncertainties about market conditions, such as the price of conventional energy. Embodying a new technology, solar-energy investments are subject to additional technological and institutional risks. These risks are in various stages of resolution.

Technological Uncertainties

The technological questions are fairly obvious: What is the lifetime of a given solar-energy system? How much conventional energy will it save over this lifetime? How reliable will the system be in delivering energy? What out-of-pocket and "hidden" costs (effort and inconvenience) are required to maintain the system? Because of local and regional variations in insolation and wind speed, the answers to these questions must be site-specific.

A potential purchaser of a solar- or wind-energy system can rarely obtain satisfactory answers to these questions. Vendors often create the impression that a particular system will save a certain percentage of utility bills. Many of these claims are mere guesses based upon unrepresentative samples for short periods of time, and most may be viewed as self-serving. Field tests currently underway at Federal and state-supported solar-energy centers can play a major role in reducing these uncertainties. But the long-term testing of existing equipment inevitably falls behind the state of the art in an environment of rapid technological change. We have only five years of operating data on equipment that was available five years ago, not on the equipment that incorporates today's new materials and concepts.

The purchaser of a particular renewable energy system may be able to transfer the technological risks to another party through a guarantee. Several lessors of industrial solar-thermal system offer contracts guaranteeing reductions in fuel bills. These guarantees are useful in sharing the risks from occasional "lemons" in a particular model. Although transferable, the underlying risk of an entire model being a lemon remains.

Until years of broad field experience can be accumulated, the best method of estimating performance is by computer simulation, the approach used here. In simulation, the user "designs" a solar or wind system on a computer as a system of equations. With actual weather data from a site, the performance of the system over any time period is simulated. But the results are only as good as the engineering assumptions.

The actual efficiency, reliability, and durability of the systems on the market today are generally less than assumed in the models (Chopra, 1980). For example, experience with the sixty-one intermediate-temperature, industrial solar-thermal systems installed since 1975 has been disappointing. One observer (K. Brown, 1981) notes that "the costs and performance of currently available solar systems have not been sufficient to create an extensive movement to install solar energy with private funds in U.S. manufacturing plants." First, only 8 to 20 percent of incident energy has been put to useful purposes. This is only one-quarter to one-half the predicted performance. Second, actual installed costs have averaged 40 percent higher than original estimates. Third, and most crucial, energy was delivered at a cost equivalent to $150/MBtu. This is far above the most generous estimate of the social cost of fuels.

It would be unreasonable to take current performance measures as representative of the future. The simulations here assume mature technology in which high levels of reliability have been achieved.

Institutional Uncertainties

The institutional environment in which solar-energy systems will operate are in the process of resolution. The major issues are the definition of sun rights; the place of solar equipment in building codes; the tax status of solar equipment; financing terms; rights and obligations of electric utilities in providing service and purchasing surplus solar power; and the warranties, guarantees, and liabilities that the manufacturer, distributor, and installer assume. These uncertainties are less formidable than they might first appear.

What guarantee does the prospective solar investor have that the sunlight will not be impeded by structures or vegetation on the neighboring property? This question is important in higher latitudes, where the sun follows a low trajectory in the sky. The conflict between sun rights and other property rights would be severe in densely populated cities with valuable land and high-rise dwellings. The courts have denied the existence of rights to unlimited access to sunlight. The landmark Fontainebleu decision (1959) upheld the right of a property owner to build a high-rise structure that shaded an adjacent swimming pool. Local land-use regulations, however, can prevent such major surprises (Klebba, 1980; Rapp, 1981, chap. 15). In high-density sites, lack of sun rights may remain a minor source of risk to prospective solar investors. In suburbs, rural areas, or low-rise commercial-industrial centers, the issue of sun rights may be academic. In those settings, each property owner may control enough land area to site collectors on roofs without risk of shading by a neighbor. By analogy, prospective developers of wind farms have had no difficulty in acquiring "wind rights" from large landholders in rural areas.

When solar-thermal and photovoltaic systems are built into inhabited structures, they are subject to building codes. These codes are administered locally, often by reference to one of several regional model codes and to nationally recognized product standards. A particular inspector may reject a given product if it does not meet the code as he interprets it. A building product has a greater probability of local approval, and hence general acceptance in the building industry, if a nationally recognized laboratory can vouch that it has met the standards for its category. A definitive review of standards finds no major obstacles to amending the codes to cover photovoltaics (Burt Hill, 1979).

The tax treatment of solar investments is currently quite favorable. The federal tax credit expires in 1985, but the possibility exists that it will be renewed. At the beginning of 1979, twenty-five states authorized or required some exemption of residential solar equipment from local property taxation, while one state provided exemptions for corporations (HUD, 1979). There is no guarantee, however, that revenue-strapped municipalities might tax solar equipment some time in the future. While tax laws can be amended or repealed, the amount of the tax credit would be known at the time of purchase and would thus not constitute a risk.

Because solar-energy systems are likely to remain expensive, most homeowners and small businesses would have to obtain mortgage financing before installing them. If the solar equipment were built into a new structure, the mortgage lender might simply bundle its costs into the value of the loan, much like built-in kitchen appliances. How solar equipment affects the terms of the mortgage depends upon how bankers evaluate its contribution to the house's resale value. Only time will tell. Financing terms would be known at the time of purchase and thus would not constitute a source of risk to the purchaser.

Several utilities are financing domestic hot-water systems. The utility leasing and maintenance of solar equipment has its precedents in the telephone industry. Because utilities are currently experiencing severe financial difficulties of their own, they are unlikely to welcome the role of financier for decentralized solar-energy systems.

For large-scale systems, examples of creative lease financing are beginning to emerge in response to the generous tax incentives for nonutility cogenerators. One motel is leasing a water-heating system from a subsidiary of a bank for a fixed monthly payment (SE, July 1979). Three textile mills are leasing thermal systems for an annual fee that is geared to conventional fuel savings (DTH, 26 August 1981, p. D-7). As noted in

Chapter 5, there are no serious financial obstacles to establishing wind farms.

The Public Utility Regulatory Policies Act (PURPA) of 1978 sketched some of the terms under which utilities will serve owners of solar-energy systems and will purchase excess electricity. As elaborated by rulings of the Federal Energy Regulatory Commission (Docket No. RM 79–55, Order No. 69), utilities must not only provide backup service to owners of solar-energy systems, but they must purchase any surplus "cogenerated" electricity at "just and reasonable rates." These rates should reflect the costs avoided by the utility by virtue of the distributed solar systems. The simulations in Chapters 5 and 6 illustrate how the avoided fuel and capital costs can be computed.

In practice, state public-utility commissions may differ widely in their rules for calculating avoided costs. At one extreme, the California commission has allowed utilities to pay cogenerators 9 to 11 cents/kwh, far above the current avoided costs of the affected utilities (ELP, July 1983). Hawaiian cogenerators are generally offered 5 cents/kwh, which is close to the avoided costs, while those in north Texas are generally offered about 2 cents/kwh, which is far below avoided costs. At the time of the investment, however, the rules can be stipulated in a contract and would not constitute a source of risk.

A pattern of warranties, guarantees, and liabilities has to be ironed out. What if the pipes of a solar-thermal system leaked, damaging an expensive painting, and the installer has gone out of business? What if a solar photovoltaic or wind system failed, burning out an appliance? Who is liable: the manufacturer, the distributor, the installer, or the consumer? The potential consumer's worst nightmare is a runaround in the midst of an electrical or plumbing crisis on a snowy Sunday morning.

In contrast to the installer who has a self-image of "businessman," the installer who is motivated by enthusiasm for a solar-energy lifestyle may lack the temperament to arise at such times to fix a customer's shorted inverter. Unfortunately, the solar industry abounds with more enthusiasts than businessmen. Undoubtedly, the emergence of solar energy as a competitive investment will create its own supply of solar businessmen. Perhaps the most comforting image would be a vertically integrated industry, with a well-known manufacturer backing up the product with long-term warranties or maintenance agreements, supported in the field with franchised installers. There are some precedents for these arrangements in kitchen appliances and air-conditioning equipment.

Economic Uncertainties

The major economic uncertainties facing the prospective purchaser of a solar-energy system are (1) the likelihood of systems being cheaper or better in the near future, and (2) the future trajectory of conventional energy prices. The possibility of cost decreases and fuel escalation necessitates evaluating solar-energy systems in real time. In other words, a system installed in 1983 is likely to have a different net present value from one installed in 1985.

Projected prices of various fuels depend upon future patterns of supply and demand. Both of these are highly uncertain and subject to unanticipated political events, such as changes in American taxes and regulations or warfare and embargoes abroad.

There have been several major efforts at modeling energy consumption and prices (EEI, 1976; Schurr et al., 1979; CONAES, 1980; Haefele, 1981). Each effort utilized alternative sets of assumptions, which generated widely diverging scenarios. Demand varies with population, economic activity, and price. Long-term forecasts (more than fifteen years ahead) of population and gross national product typically err by about 10 percent; energy demand forecasts err by even more (Ascher, 1978, chaps. 3–5). There is also substantial uncertainty about how responsive user demand is to changes in price. The supply picture is hardly clearer. Nevertheless, a future of increasing pressure on energy resources and rising prices emerges under all plausible scenarios.

Several authoritative studies projected oil-price increases in the range of 1.2 to 3.7 percent per year for the period from 1975 to the year 2000. Nordhaus (1979, Table 6-5) projected year 2000 crude oil prices of $1.50/MBtu in 1975 dollars, while other studies he reviews project prices as high as $2.85/MBtu. Edison Electric Institute's (1976, Table 6-79) scenarios were in the range of $2.00 to 2.60/MBtu. Oak Ridge National Laboratory (Hirst and Carney, 1977) projected oil prices to rise to $3.75/MBtu by the year 2000. The Ford Foundation energy study (Landsberg et al., 1979, Fig. 2-8) projected the highest prices, $4 to 5/MBtu. The track record of these projections is rather poor. By 1978 the world price of oil had surpassed $18/bbl. (or $3.00/MBtu), which translates into $2.50/MBtu in 1975 prices. The plummeting of Iranian production following the revolution resulted in a rapid doubling of world oil prices, to about $35/bbl. ($6/MBtu), or $4.50/MBtu in 1975 prices. The ensuing decline in world prices to $30/bbl. in 1983 brought costs back down to $3/MBtu in 1975 dollars.

Political events will surely result in perturbations in world oil markets in the next twenty years. Difficulties in both economic and political forecasting notwithstanding, the price of oil is likely to escalate at the annual rate of 2 to 4 percent above inflation into the foreseeable future. The long-term escalation rate is taken as an abstraction, for there are likely to be abrupt rises in oil prices, followed by periods of stability and even decline, which may offer the illusion of an oil glut. The theoretical basis for selecting escalation rates is considered in the Appendix. It is not unreasonable to expect natural gas prices to escalate at the same rate as oil, its main competitor.

Because of its relative abundance, coal may rise more slowly in price than oil and gas. Other factors may cause coal prices to follow those of oil and gas. These include unforeseen tightening of environmental laws regulating land reclamation, air emissions, and waste disposal, or a tightening demand as an international market for coal develops.

As the prices of oil and gas continue to rise, synthetic liquids and gases derived from coal will become increasingly economical. Technically, liquid synthetics could satisfy the entire gamut of energy demands. The cost of these synfuels will reflect both the cost of coal, which will continue to escalate, and the cost of capital, which is difficult to forecast. In the 1970s the cost index for items used in large-scale energy facilities rose faster than consumer prices (EEI, 1980, Tables 60 and 62). Whether this trend continues is difficult to ascertain. Capital costs may even fall as the synfuels industry moves down its learning curve. The price of synfuels may rise more slowly than that of coal, and it may provide a ceiling beyond which oil and gas prices cannot rise. Ultimately, the cost of fuels produced by this expensive backstop technology may become the standard against which solar-energy technologies will be evaluated.

A recent study by Resources for the Future (Schurr et al., 1979, p. 54) estimated that synfuels could be produced at twice the 1977 prices of natural gas and petroleum, or $28/bbl. ($31/bbl. in 1980 costs). World prices have already reached this mark, yet efforts of private enterprise to develop a synfuels industry without government subsidies (price or loan guarantees) is rather tentative. This reluctance may be symptomatic of the chronic tendency of analysts to underestimate the costs of producing synfuels. Construction of the world's largest synfuels project, which was to extract oil from shale, was recently canceled after experiencing cost overruns as high as 60 percent (BW, 19 April 1982). A definitive review of synfuels (Harlan, 1982) suggests the plausibility of synfuels costing as much as $80/bbl. Like a mirage, the economic viability of synfuels always appears just out of reach. Because synfuels are likely to cost at least as much as natural oil or gas, separate projections of synfuels prices are ignored in the remainder of this analysis.

Dealing with Risk: The Example of Domestic Hot-Water Systems

All approaches to dealing with risky investments begin with laying out scenarios that that can affect costs and benefits. Each scenario embodies a particular set of contingencies. For example, one might assume that a solar-energy system will perform perfectly 99 percent of the time, that utilities will purchase electricity for the value of the fuel saved, and that fuel prices will escalate at 2 percent a year. Once the assumptions for any scenario have been stated, one can look at the bottom line: What is the value of the conventional energy saved, net of system costs? This question has many variants. What is the best size of array and storage? Is it desirable to wait until system prices come down? Does the value of electricity sold back to the utility compensate for the cost of the two-way hookup?

Because there are so many plausible scenarios, some form of summary would be helpful. If probabilities could be ascribed to each scenario, or at least to the assumptions underlying each scenario, then an "expected" value could be calculated. For example, one might believe that there was a 20 percent probability of fuel escalating at 1 percent, a 60 percent probability of escalating at 2 percent, and a 20 percent probability of escalating at 3 percent. At 1 percent escalation, the solar system might provide a present value of −$1,000; at 2 percent, a present value of $2,000; and at 3 percent, a present value of $3,000. At the assumed probabilities, the expected net present value of the system is $1,600. When large numbers of contingencies (such as fuel escalation rates and maintenance costs) are dealt with together, the distribution of net present values are useful as well. An analyst may want to know, for example, the probability that the present value of the investment is negative.

The calculation of expected values is unworkable in the case of solar technologies because of the difficulty of assigning probabilities to political events, such as a change in PURPA or an oil embargo, or technological innovations, such as a breakthrough in synfuels production or photovoltaics.

If the underlying probabilities are unknown, one can assume that all of the plausible scenarios are equiprobable and that the contingencies affecting costs and returns, such as maintenance costs and fuel-escalation rates, are independently distributed. A conservative investor may be interested in the consequences of the worst-case combination of contingencies, whereas a "wildcatter" might be interested in the best-case combination. Most investors would be less interested in the extremes, which are rare events, than in the median outcome and perhaps the outcomes in the top and bottom deciles.

Some Solar Scenarios

The steps in assessing the prospects for a solar-energy system include assaying the availability of solar energy, simulating the performance of the system, and evaluating the energy it produces.

Estimates of insolation in the United States are based upon measurements obtained for nearly thirty years from more than two dozen weather stations. While many of the reported data appear to be erroneous, the sampling points paint an adequate picture of broad regional patterns (Rapp, 1981, pp. 75–85). Annual insolation generally decreases as one moves from the arid Southwest to the humid Northeast.

A 40-square-foot solar-thermal system for domestic hot water costs about $2,000. Computer simulations indicate that this system can save annually about 11.5 MBtu in Los Angeles, 9.7 MBtu in Atlanta, and 7.0 MBtu in Indianapolis (SE, July 1980). These cities are representative of good insolation conditions in the semiarid Southwest, the humid Southeast, and the humid North, respectively. Conventionally, this energy may be delivered by electric-resistance heating or by the direct combustion of oil or gas.

As a first approximation, regional variations in energy prices are ignored and national market prices used: $18/MBtu for electricity, $9/MBtu for home-heating oil, and $4.50/MBtu for residential natural gas (MER, March 1983). As noted, electricity prices underestimate social costs by the exclusion of externalities, and natural gas prices are controlled below their replacement cost.

In comparison to an electric-resistance water heater, a solar-energy system yields the typical homeowner a gross annual savings of $210 in the Southwest, $175 in the Southeast, and $125 in the North. Compared to an oil heater, the solar savings are only half as much: $104, $87, and $63, respectively. Compared with a natural-gas heater, the solar savings are one-quarter as much, or only $52, $44, and $32, respectively.

In addition to these initial capital costs are the recurrent costs of maintenance. These include minor repairs of corroded or broken parts and replacement of fluids or major components on a regular cycle. In addition, some electricity is required for operating the pump.[4] Maintenance and operation costs are not well known, but they are probably 1 to 2 percent of the initial system costs, or $20 to $40 annually (SE, July 1980, pp. 10–11; SEC, March 1982, p. 23). These annual maintenance costs are not trivial when compared to annual energy savings. An annual $40 maintenance bill would practically cancel any savings over gas heaters.

The durability of solar-thermal systems depends somewhat upon the quality of maintenance. The lifetime of these systems has been estimated to be in the range of fifteen to twenty-five years. At the end of its lifetime, the solar-energy system must be disposed of and parts salvaged. The disposal cost and salvage value are probably small and may even cancel out each other.

For purposes of illustration, we follow a middle range scenario that assumes that the solar hot-water system lasts for twenty years, with

32 Solar and Wind Energy

maintenance cost of 1.5 percent of capital costs ($30 a year). Three common criteria for comparing the costs and benefits or investments are: (1) payback period; (2) net present value; and (3) levelized costs and benefits.

Payback Period

The payback period is the number of years after which the cumulated benefits (energy saved) equal the initial cost of the solar-energy system. Program PAYBACK in the Appendix calculates the payback period with and without tax credits and at various fuel-escalation rates.

As a benchmark, let's consider the payback period in the absence of tax credits and any fuel escalation. When compared to an electric-resistance heater, the solar-energy system gives a payback period of eleven years in the Southwest, thirteen years in the Southeast, and nineteen years in the North. When compared to both oil and gas heaters, the payback period in all three regions exceeds twenty years, the expected system life *(Table 2.1)*.

The annual escalation of conventional energy prices by 4 percent above inflation reduces the payback period by up to five years when solar systems are compared to resistance heating. The payback period for the solar-energy system is close to its lifetime when compared to oil heaters and still exceeds its lifetime when compared to gas heaters.

The 40 percent federal tax credit reduces the payback period by 40 percent. Additional tax credits available in several states reduce the cost and payback period even further. Even with both the federal tax credit and 4 percent fuel escalation, the payback periods of solar heaters are fairly long. Compared to oil heaters, solar-energy systems enjoy a twelve-year payback in the Southwest, a fourteen-year payback in the Southeast, and an eighteen-year payback in the North. The payback period exceeds the life of the solar system when compared with gas. If deregulation raised the gas price to that of oil, the payback period would become less than the lifetime of the system only because of the federal tax credit.

Net Present Value

A solar-thermal system can be evaluated by comparing the capitalized benefits with the initial system costs. Algebraic equations and Program PW SAVINGS for computing the net present value of solar energy systems are presented in the Appendix.

For illustration, the net annual savings of solar thermal systems are evaluated at $18/MBtu, $9/MBtu, and $4.50/MBtu. The savings are

Table 2.1 Payback Period of Solar Domestic Hot-Water System

	Conventional Energy Source		
no tax credit no escalation	electricity	oil	gas
Southwest	11	24*	63*
Southeast	13	30*	85*
North	19	47*	177*
no tax credit 4% escalation			
Southwest	9	17	28*
Southeast	11	19	32*
North	14	25*	39*
40% credit 4% escalation			
Southwest	6	12	21*
Southeast	7	14	24*
North	10	18	31*

*Exceeds lifetime of system.

Source: Energy savings from "Wave the Flag for Solar Savings," *Solar Engineering* 5 (July 1980):10–11.
  Savings for solar DHW normalized for 40-square-foot system: 11.49 MBtu in Los Angeles, 9.68 in Atlanta, 6.96 in Indianapolis.

escalated and discounted under three scenarios: worst case (10 percent discount/2 percent fuel escalation); most likely (8 percent discount/ 3 percent fuel escalation); and best case (5 percent discount/4 percent fuel escalation). When these savings are evaluated at $9/MBtu, the value of savings is far below the social cost of the systems ($2,000) under all scenarios. When the 40 percent federal residential tax credit is taken into consideration, the value of oil savings exceeds the costs ($1,200) in the two sunny regions under the best-case scenario. Even under the latter scenario, tax credits are insufficient to render solar-thermal systems competitive with conventional energy priced at $4.50/MBtu.

When compared to an electric hot-water heater, the present value of the savings from a solar system is below its social cost under the worst-case scenario in all three regions. Under the most likely scenario, the savings exceed social costs in the Southwest, but not the Southeast.

Table 2.2 Net Present Value of Savings of Solar Domestic Hot-Water System

	Value of Conventional Energy ($/MBtu)		
Worst Case 10% discount/2% fuel	$18 electricity	$9 oil	$4.50 gas
Southwest	1801*	713**	259**
Southeast	1479*	612**	176**
North	996**	370**	58**
Most Likely 8% discount/3% fuel			
Southwest	2317	1011**	358**
Southeast	1908*	807**	256**
North	1295*	500**	103**
Best Case 5% discount/4% fuel			
Southwest	3376	1501*	564**
Southeast	2789	1208*	417**
North	1909*	767**	597**

*Below cost of system without tax credit.
**Below cost of system with 40 percent federal tax credit.

Under the best-case scenario, savings exceed social costs in the two sunny regions, but not quite in the Northeast. When costs are reduced by the federal tax credit to $1,200, the solar-energy system becomes competitive in the Northeast in the more favorable scenarios (*Table 2.2*).

Bringing the market prices of conventional energy up to their social costs makes domestic hot-water systems more attractive. If electricity rates were increased by about 10 to 20 percent to reflect externalities, these systems would be attractive in both southern regions even without a tax credit. The national-security premium on home-heating oil would have to approximate 100 percent to render these systems competitive without a tax credit. The deregulation of natural gas to $7.50/MBtu, plus a 25 percent national-security premium, would still leave solar heating uncompetitive with gas heating.

Levelizing and Its Pitfalls

The utility industry's criterion for investment analysis has particular value in rate hearings. Instead of discounting benefits to the present, utilities artificially spread out the costs of a project evenly over its lifetime. When discounted back to the present, "levelized costs" equal the present value of costs. These costs indicate the average fees for service that the utility must charge in order to recoup all of its costs, including fuels, maintenance, taxes, repayment of principal, and dividends or interest. The investment that produces electricity at the lowest levelized cost is to be preferred. A common example of levelization is a flat mortgage payment, which distributes a lump-sum initial cost as if it were a monthly stream. Algebraic equations and Program LEVELIZATION are in the Appendix.

To illustrate, the levelized capital cost of the $2,000 solar hot-water system is $204 when the interest rate is 8 percent. Adding the $30 maintenance and operation, the levelized social cost is $234. Dividing $234 by the energy delivered annually, the system costs $20/MBtu in the Southwest, $24/MBtu in the Southeast, and $33/MBtu in the North. These costs lie above current costs of electricity, which average $18/MBtu.

Although levelization and capitalization lead to similar conclusions, levelization has several disadvantages. First, levelization is no longer adequate in projecting future energy costs. This practice made sense in the pre-1970 period, when the underlying cost structure was stable. In an era of rising fuel and construction costs, the face validity of levelization diminishes. Levelized electricity costs correspond to actual electricity charges at neither the beginning of the investment period nor the end, but in some year in between. For purposes of comparison, conventional energy costs must be levelized at some assumed rate of escalation. For example, at 8 percent interest, the levelized cost of electricity, beginning at $18/MBtu. and escalating at 3 percent per year, is $23/MBtu. This is still below the $24/MBtu. cost of solar hot-water heating in the Southeast.

Second, the computed levelized cost of solar systems may create the impression that the costs, benefits, or loan repayments for the solar-energy system are in fact or should be level. There is no reason for this to be so. While any stream of benefits or costs has a unique capitalized value at a given discount, rate the same capitalized value is consistent with many streams of benefits and costs. This fact enables financial intermediaries like savings banks to design innovative schemes for financing solar-energy systems that match the time path of loan repayments to that of benefits.

Third, participants in capital markets are experienced in computing discounted present values. Assets are bought and sold at their capitalized, not their levelized, values. In the hypothetical example, the net present value of the solar system in Los Angeles under one scenario was

$1,100 ($2,300 minus $1,200). This represents the premium above costs that a homeowner with an resistance heater would be willing to pay. Levelized values provide no such information.

Solar Energy as Insurance

The scenarios indicate that the major contingencies that affect the economics of solar-energy systems at a given site are: (1) technological performance relative to specifications; (2) longevity of the system; (3) maintenance costs as a proportion of initial capital cost; (4) the fuel escalation rate; and (5) the discount rate. The first three items refer to technological contingencies, the last two to economic or market contingencies. The following numbers appear to represent the worst, most likely, and best contingencies for each, respectively:

Performance: 90, 100, and 110 percent of specifications
Longevity: 15, 20, and 25 years
Maintenance: 2, 1.5, and 1 percent of capital cost
Annual fuel escalation: 2, 3, and 4 percent
Discount rate: 10, 8, and 5 percent

The elaboration of a large number of scenarios hardly constitutes a complete risk analysis. At a minimum, two other characteristics of these scenarios are important. First, what is the shape of the distribution of outcomes from these scenarios? For example, one may want to calculate the probability that the outcomes will be positive. Second, what is the relationship between the outcomes of solar-energy systems and outcomes of other components of the "energy portfolio," especially conventional energy costs? For example, if solar systems yield the highest return when conventional energy costs are high, then solar energy constitutes valuable insurance.

To explore these issues, let's compare solar hot-water heating to electric heating in the Southwest. The solar system is expected to reduce energy consumption from 14.4 MBtu annually to 2.9 MBtu, at a value of $18/MBtu. In the absence of any knowledge of the underlying probabilities, the contingencies are taken as equiprobable and independent. The five sets of contingencies generate 243 possible scenarios. Under each scenario, three present worths are calculated: (1) net solar savings; (2) no-solar energy costs; and (3) energy costs with the solar system. The latter sums the costs of the solar system and the residual electricity bill (*Table 2.3*).

The distribution of net solar savings is positively skewed, with a mean of about $425 and a median of $280, which straddle the result from the most likely scenario—about $320. The extreme scenarios yield returns

Table 2.3 Net Present Worth of Solar Savings, No-Solar Energy Costs, Energy Costs with Solar: Scenario Distributions

	Mean	Stand. dev.	Scenarios			
			Market risks only		All risks	
			discount = 10% fuel = 2%	discount = 5% fuel = 4%	discount = 10% fuel = 2%	discount = 5% fuel = 4%
Solar savings	426	669	−199	1376	−694	2757
No-solar energy cost	3495	866	2632	4801	2291	5864
Net energy costs w/solar	3069	347	2831	3425	2984	3107

Beta (solar savings vs. no-solar) = −.72
Beta (net energy w/solar vs. no-solar) = .28

of −$690 and $2,750. If there were no technological risks, the range would be far less, −$199 and $1,376. Interestingly, about 15 percent of the cases lie below the worst-case scenario, and about 10 percent lie above the best-case scenarios. In other words, the range between the worst and best cases as specified by combinations of discount and fuel-escalation rates spans about three-quarters of the plausible outcomes. In the chapters that follow, then, using such worst-, most likely-, and best-case scenarios appears to be a reasonable treatment of risk.

The variance in the net present value of solar savings can be factored into technological risks (performance, longevity, and maintenance) and market risks (fuel-escalation and discount rates). Nearly 90 percent of this variance can be attributed to market factors. This finding suggests that the large technological uncertainties postulated here are relatively unimportant.

The no-solar energy costs have a relatively high standard deviation, which indicates that being a consumer of conventional energy is risky. The variations in net solar-energy savings are somewhat less. More interesting, the variation in the energy bill with solar is far below the variation in either the no-solar costs or solar savings. This is because solar-energy systems reduce energy costs the most when conventional energy costs are highest. In other words, the best-case scenario for solar energy is the worst-case scenario for the consumer who does not have a solar-energy system. Because a solar-energy system enables a consumer to diversify his energy "portfolio," solar energy constitutes an insurance policy.

The value of this solar-energy system as an insurance option is indicated by *beta*. A perfect insurance instrument would completely offset risks and have a *beta* value of −1.00. The domestic hot-water system in this case has a *beta* value of −.72. The insurance characteristic of solar energy has two implications. First, risk-reducing assets like insurance are desirable in some cases even though the expected value of premiums exceeds the expected value of returns. Hence, a solar-energy system might be desirable even if the net present value of its savings were slightly below the net present value of its costs. Second, financial markets generally discount less risky investments (those with low *betas*) at relatively low rates. This suggests that the 5 percent discount rate may be more relevant than the 10 percent discount rate.[5]

Conclusion

Because of distortions in energy markets, the social value of conventional energy displaced by solar systems is greater than the value calculated by the homeowner or business executive. While the tax system has a major impact on the evaluation of solar-energy systems, it does not fully correct for these distortions from the viewpoint of business investors. Generous tax credits for homeowners, however, may make solar investments appear profitable even when the costs exceed the social value of the conventional energy saved.

Just because there are significant technological, institutional, and economic uncertainties surrounding solar-energy systems does not imply that solar investments are risky. If good "most likely" estimates of technological performance can be specified, solar investments can be viewed as insurance assets. This insurance value is a visible benefit to the investor in terms of reduced variability of total energy costs, and it can be added to the net present value of energy savings.

Insurance benefits aside, the example of domestic solar hot-water systems indicates that they are competitive with electric heating only in the sunniest locations. In most locations, these systems are not competitive with home-heating oil or natural gas, even under scenarios of high fuel-escalation and low discount rates. Such systems are far from competitive with oil or gas heating in the Northeast and Southeast. Competitiveness can come only from significant decreases in system costs. This is even more clearly the case for photovoltaic systems. But what are the prospects for costs falling sufficiently?

Notes

1. While government has been credited with taking the long view, it often behaves as if its time horizon were two to four years, when another election looms. For a review of the vagaries of American energy policy, see Goodwin (1981).

2. Lind (1983b) suggests that the price of investment be 3.8 and the price of consumption 1.56. For investments financed by reducing consumption, both costs and benefits should be multiplied by the same 1.56 factor and then discounted by the social rate of discount of 4.6 percent. In this case, corrections cancel. For investments financed by reducing private investment, multiply costs by 3.8 and benefits by 1.6 and then discount them by 4.6 percent. These latter corrections are equivalent to discounting a twenty-year stream of benefits at 18 percent per year. The capital-intensive nature of solar technologies suggests that they will have to be financed by some reduction in consumption and increase in savings. Assuming that solar investments are financed in equal proportions by reduced investment elsewhere and by reduced consumption, the correction factor for costs becomes 2.8. This correction is equivalent to discounting a twenty-year stream of benefits at 10 percent.

3. Once there is sufficient experience with resale of solar homes, the resale value of solar systems can be easily computed. The technique of computing "hedonic price indices" involves regressing housing prices against housing characteristics, such as the presence or absence of solar systems. See, for example, Martin T. Katzman, "The Contribution of Crime to Urban Decline," *Urban Studies* 17 (October 1980) : 277–86.

4. In common domestic hot-water systems, pumps are required to move the hot water from the rooftop arrays to a storage tank in a garage or basement. The thermosiphon

system, commonly used in Israel, exploits the natural tendency of heat to rise by placing the storage tank above the arrays on the rooftop. Aesthetic objections aside, few existing roofs have been designed to support fifty-gallon tanks. At some cost, these aesthetic and structural problems might be overcome in new American dwellings.

5. Lind (1983b) suggests that with a *beta* of minus one the correct social discount rate should be 3 percent. The present value of a twenty-year stream of benefits discounted at 3 percent is 20 percent higher than a stream discounted at 5 percent. If Lind's argument is accepted, then our calculations undervalue the gross savings of solar-energy systems by 20 percent.

3
Bringing Down the Costs

The widespread commercial adoption of solar-energy systems depends critically upon significant technological progress. Progress may take two forms. One is a cost-reducing improvement in the production process. Another is the development of a new engineering concept that improves product performance. Lowering the cost of a given solar-thermal system from $2,000 to $1,000 exemplifies process improvement, while increasing the efficiency of thermal conversion exemplifies product improvement. Both manifestations of technological progress reduce the cost per unit of energy delivered. For widespread adoption in most parts of the country, the cost of solar hot-water systems must be reduced by a factor of two from 1980 prices. For solar-thermal space heating, the required factor is four, and for photovoltaics the factor is ten. The required reductions would be even larger if federal tax credits were eliminated. What are the chances of such major improvements?

Progress in Existing Technologies

Economies of Scale

In a new industry considerable room exists for cost reductions or performance improvements. Process improvement can occur if there are significant unexploited gains to the division and specialization of labor and machinery. When the market is small, the division of labor is uneconomical, but when the market grows, the division of labor enables firms and industries to enjoy economies of mass production, or economies of scale. Since economies of scale depend upon the annual flow of output, they are reversible. If the demand contracts and production falls, costs may rise. Some production technologies lend themselves to quite substantial economies of scale, so only a few firms can achieve the most economical size. This may be the case in the photovoltaic industry, as indicated by the potential economies of scale in various steps in module fabrication.

The possibilities that a successful innovator may capture a large market share can be a positive stimulus to privately funded basic research on photovoltaics. Because basic research may produce concepts that are easily copied by a competitor, private firms are often reluctant to engage in such research. The prospect of monopolizing a market by the earlier embodiment of superior concepts in equipment can offset this risk.

The Industrial Learning Curve

Improvements in the performance of innovative technologies over time have been observed in many contexts (Arrow, 1962; Lenz, 1968). This phenomenon has been described as the industrial learning curve. Continuous reductions in unit cost or improvements in performance can emerge from the familiarization of the operatives, technical staff, and management with the production process. While firms are not organic entities capable of true learning, persons in an organization can translate their learning into improvements in the process of manufacturing a given product, or into engineering and design improvements in the product itself.

Learning on the part of the industry can occur through technical progress of common suppliers, imitation of competitors, and mobility of labor and management from firm to firm. Large electronics firms, for example, have been incubators of much entrepreneurial and managerial talent in the nascent photovoltaic industry (SE, February 1980).

Industrial learning is the result of *cumulative* production experience. Unlike the economy-of-scale phenomenon, industrial learning is not easily reversed if the rate of annual production declines. If production and employment are cut back, the skills of the labor force deteriorate slowly and are easily reacquired.

Industrial learning curves can be described by the mathematical formulas (Fusfeld, 1973):

$$C = cN^{-a} \quad 0 < a < 1$$
$$P = pN^b \quad 0 < b < 1$$

N is the cumulative volume of production. The rate of cost reduction a and performance improvement b is thus a function of the rate of growth of output:

$$\ln C = \ln c - a \ln N$$
$$dC/C = -a \, dN/N$$

And similarly:

$$\ln P = \ln p + b \ln N$$
$$dP/P = b \, dN/N$$

If output grows at a constant annual rate k, then $N_t = N_0 e^{kt}$, and $\ln N_t = \ln N_0 + kt$ and $dN/N = k$

Combining the first three equations, we obtain the traditional trend forecast:

$$dC/C = -ak$$
$$dP/P = bk$$

In a classic article on industrial learning curves, Arrow (1962) reviewed several studies that identified progress parameters (a, b) on the order of .3 to .5. The capacity of a firm to learn from experience appears to vary directly with the average educational and experience levels of the work force and the rate of investment in productive capacity. These magnitudes indicate that a doubling of cumulative output results in a diminution of per-unit prices by 30 to 50 percent.

The industrial learning curve for silicon transistors is particularly germane to the solar photovoltaic industry. From 1953 to 1959 progress was relatively slow. As cumulative volume increased from .015 million units to 15 million units (at an average annual growth rate of 350 percent), unit costs declined along a 10 percent slope, implying a .1 progress parameter. From 1959 to 1964 progress accelerated. As cumulative output increased from 15 million to 450 million (at an average annual growth rate of 100 percent), progress exceeded the 30 percent slope, implying a progress parameter larger than .3. After 1964 the rate of progress followed the 20 percent slope. Over the entire fifteen-year period, from 1953 to 1968, the price of transistors decreased from $15 to 40 cents per unit (a 40-fold reduction). While the progress parameter averaged .23 (*Figure 3.1*), it is clear that technology does not progress uniformly but in fits and starts.[1]

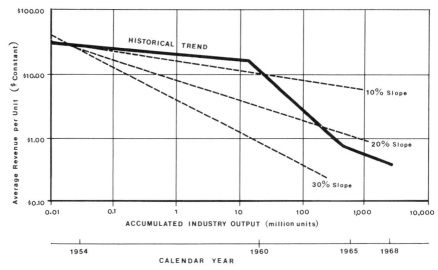

Figure 3.1 The learning curve for silicon transistors. Source: after Palz (1978), p. 215.

The Inducement of Energy Innovations

The anticipation of continuing increases in energy prices elicits different responses in the short and long run. In the short run, society will try to conserve energy with the equipment currently used. In the long run, it will attempt to develop better equipment.

In the short run, a wide range of existing equipment is available to perform a given function, say heating fifty gallons of water to 120 degrees F. This equipment varies in capital cost and energy efficiency. These existing technological options can be represented by an isoquant, a curve of equal outputs. For example, I_0 represents the many technologies for heating fifty gallons available in 1970 (*Figure 3.2*). The slope of P_0P_0 is the contemporary relative price of capital and energy. Under these conditions, point A_0 is the least-cost capital equipment/energy combination. An increase in energy prices is represented by an increase in the slope of the price line from P_0P_0 to P_1P_1. In response to the energy-price increase, the least-cost capital/energy combination changes from A_0 to B_0 along isoquant I_0. When it is time to replace their equipment, consumers will purchase model B_0 instead of model A_0.

The state of science and engineering offers a range of potential production technologies much wider than those actually exploited. The realization of a potential technology is dependent upon investments in

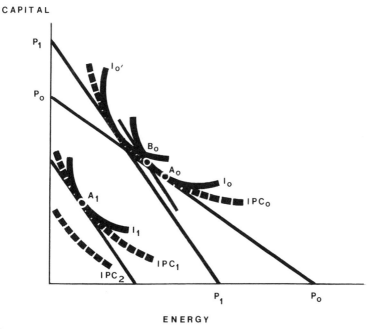
Figure 3.2 Innovation possibilities frontier and achieved isoquants.

research and development, mostly by suppliers to the producing industry (Binswanger and Ruttan, 1977). For example, the equipment represented by all the combinations on curve IPC_0 could have been developed from the technological base of 1970. From this innovation-possibilities curve, a set of technologies was selected for development that minimizes costs to consumers (and maximizes profits of their suppliers) at the given factor prices. The technologies represented by I_0 were developed when relative energy prices were expected to be P_0P_0, but I_0', which was neglected, would have been more profitable in retrospect.

In the interim, the state of science and engineering improves innovation possibilities for a given level of research and development expenditures. For the same level of expenditures as before, a supplier can now develop technologies on the innovation possibilities curve IPC_1. For a higher level of expenditures, a superior set of innovation possibilities presents itself, IPC_1'. Under the new technological possibilities and price expectations, suppliers will strive to achieve a technology A_1. While all may compete to achieve such a technology, some may fall short of the mark and achieve other points in isoquant I_1. How much suppliers will devote to the development of a particular realm of technology depends upon the economic payoffs. These depend upon the relative cost reductions achievable in their various product lines and the size of the market.

In the arena of energy-using equipment, suppliers may compare the payoffs of developing more energy-efficient models (a conservation/demand-oriented strategy) versus developing new sources of energy (a supply-oriented strategy). In other words, the development of technologies embodied in more efficient heat pumps or gas heaters competes for research dollars with the development of renewable technologies. A given R&D budget may reduce the cost of heat pumps and photovoltaics by the same 10 percent, but the gain in sales resulting from the former may dwarf the gain in the latter. This market reality may impede the actualization of technological possibilities.

Prospects for Solar-Energy Systems

It is difficult to forecast the possibilities for technological progress in any field. Clearly, they depend upon the maturity of the underlying science. Energy-conversion systems based on well-established fields like thermodynamics, fluid mechanics, or aerodynamics are less likely to benefit from scientific breakthroughs than systems based upon genetics, photochemistry, or electronics. Engineering breakthroughs in materials cannot be ruled out for any system of solar-energy conversion. For example, new metallic alloys and types of fiberglass are being tested in experimental wind-energy conversion systems.

The prospects for substantial cost reductions or performance increases in solar-thermal systems are not that bright. Scientific advances in the field of solar-thermal energy conversion are unlikely. Breakthroughs in engineering concepts or new materials appear to be elusive. Despite a considerable injection of funds from large corporations, technological progress in solar-thermal systems seems to have bottomed out. The most optimistic projections are for cost decreases of about 15 percent for domestic hot-water systems (Jacobsen and Ackerman, 1981). Several oil companies and well-established materials suppliers have begun divesting themselves of solar-thermal subsidiaries, while redeploying their resources into photovoltaics. Many independent fabricators of solar-thermal systems share this brighter view of photovoltaics (SE, December 1980 and June 1981).

A major limitation to cost reduction is the large amount of labor required for installation. In existing homes, installation is nearly a customized process. The potential for labor-saving modularization appears greater for installations in tract housing or multifamily dwellings. Thus far, there is too little experience with mass installation to draw any conclusions about its economic potential.

In 1976 the Department of Energy's National Photovoltaic Plan envisioned a significant reduction in array costs per peak kilowatt: from $21,000 in 1976 (in 1980 dollars) to $6,000 in 1980 to $700 by 1986. Complete PV systems were projected to cost $1,600 per kilowatt in 1986. These "projections" were really targets rather than forecasts of likely events. At the targeted costs, photovoltaics should be competitive in a wide range of commercial applications. How can the likelihood of realizing these targets be assessed?

Technological Forecasting

Technological forecasting is the assessment of the likelihood of major cost reductions or performance improvements. In general, forecasters have attempted to foresee continuous improvements in existing processes and products as well as discontinuous breakthroughs, or revolutions. Technological forecasts have focused upon the following events: the demonstration of a new technological capability, the successful field test of a prototype, initial commercial adoption, widespread diffusion, and the social consequences of an innovation (Bright, 1973).

In the case of photovoltaics, technical capability was demonstrated in the 1950s, extraterrestrial applications were begun in the 1960s, and a program of field testing was initiated in the 1970s. Studies of the diffusion of innovations have demonstrated that the timing of initial adoption, rate of diffusion, and ultimate market penetration are a function of such factors as profitability, size of initial commitment, and rate of replacement and addition to the capital stock. In turn, potential producers of innovations devote their inventive and productive efforts to those applications and locations where they perceive the market to be largest (Griliches, 1960; Mansfield, 1971; Schmookler, 1971). If the perceived market is small or unprofitable, producers will devote few resources to new product development, and a potential innovation may not materialize.

The purpose of forecasting innovations should be made explicit at the outset. It is useful to forecast events that are beyond the control of the audience for which the forecast is made but that have great impact on their activities. For example, homebuilders would be interested in the relative costs of photovoltaics and conventional energy systems so that they could anticipate the demand for solar homes. It is less useful to forecast events that are under the control of the audience to which the forecast is directed or to forecast an event that one hopes to bring about. Indeed, if public policy aims to induce the solar industry to achieve certain cost goals, it is meaningless to forecast cost as if it were given. Such forecasts can even be self-defeating. Instead, forecasts should be

48   *Solar and Wind Energy*

seen as consequences of specific actions, such as greater expenditures for basic research or consumer education.

Of the many methods of technological forecasting, two have been applied to photovoltaics: curve fitting and the Delphi Method.

The Photovoltaic Progress Function

Few observers forecast major breakthroughs in the efficiency with which silicon photovoltaic cells convert sunlight into electricity. Indeed, the major breakthroughs in efficiency occurred in the 1950s and early 1960s. Daniels (1964) notes that demonstrated silicon-cell efficiencies were 1 percent in 1953, 11 percent in 1955, and 14 percent by 1960. While somewhat higher efficiencies have been achieved in the laboratory, cells on the market today have efficiencies of about 11 percent. Since the theoretical maximum efficiency of silicon conversion is 23 percent, at best a doubling of efficiency can be expected.

The actual cost history of silicon cells indicates extremely rapid technological progress. Palz (1978) estimates annual worldwide production at 10 kw per year in the period 1958–73. Constant annual output means that the percentage growth of cumulative output is *decreasing*. Since then, output has grown more than 100 percent per year, largely as a result of government research and development programs.[2] In current dollars, arrays cost $500,000/kw in 1958 and $15,000 in 1975, when the Department of Energy (then the Energy Research and Development Administration) established its ten-year goals. In the intervening eighteen-year period, consumer prices doubled, so real costs fell by a factor of over 65, or at annual rate of 26 percent. The achievement of the Department of Energy goals implies an annual cost reduction of 40 percent, which is faster by half than prior experience (*Figure 3.3*).

The photovoltaic progress must be interpreted with qualification. First, the early portion of the curve describes extraterrestrial applications, while the latter portion describes earthbound applications. Since the former cells require less elaborate and hence less expensive encapsulation, the curve understates the rate of cost reduction. The curve also aggregates purchases of different magnitudes. Since larger orders can be filled at a lower unit cost than smaller orders, the curve mixes the static economies-of-scale effect with the dynamic learning-curve effect.

The annual cost reduction is a product of the technical progress parameter and the annual rate of growth of output. Continuous annual cost reductions of 40 percent can be achieved by many combinations of the terms. Two examples: (1) a low technical progress parameter (.1) and a high (400 percent) annual growth of output, which was experienced in

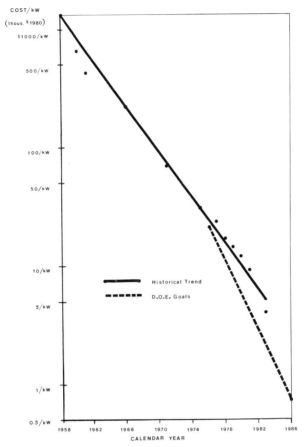

Figure 3.3   The learning curve for photovoltaics. Sources: Prince (1963); Ul-Rahman (1967); Sheridan (1972); Moore (1976); Maycock (1978); Murray (1981); PI (May 1983).

the early 1960s; or (2) a high technical progress parameter (.3) and a moderate (133 percent) annual growth of output.

Although the 40 percent rate of cost reduction is arithmetically possible, the achievement of this rate is hardly plausible. First, few industries, including the silicon transistor industry, have experienced such dramatic cost reductions. Second, the output of the photovoltaic industry has never grown as fast as 100 percent. Third, actual photovoltaic cost reductions during the 1976–81 period have been closer to the 26 percent historical trend than to the targeted 40 percent rate. Fourth, technological forecasts have been conspicuously wrong in the past (Ascher, 1979). Few technological forecasters have considered the cost trends of innovations. Instead, they have focused mainly upon technical capabilities. The most conspicuous example has been forecasts of computer capabilities, such as the number of computations per second. Five-year forecasts made in the late 1960s, after two generations of computers had been developed, were on the average 60 percent off the mark in both directions. For a less developed technology like photovoltaics, larger errors of forecasting can be expected.

The learning curve describes the performance of the new technology in its innovative range. Since progress eventually tapers off, forecasters identify a minimum level of cost or a maximum level of performance that is achievable with a given technology. Progress in the performance of a general function, such as the direct conversion of sunlight into electricity, can continue by switching to a new technology. When the new technology attains its limits, another may be invented to perform the same function. Forecasters thus focus upon the derivation of envelopes of functionally equivalent technologies (Ayres, 1968).

In the case of photovoltaics, silicon-cell flat-plate arrays may be superseded by silicon-cell concentrator systems, which use parabolic mirrors or Fresnel lenses to magnify the photon influx on the cells. Polycrystalline silicon may give way to more exotic materials. Promising alternatives include more expensive gallium arsenide, which has a higher photoelectric-conversion efficiency, and cheaper amorphous silicon, which converts at a lower efficiency. The high cost and efficiency of gallium arsenide cells may recommend their use in concentrating systems, where the major capital costs are embodied in the mirrors. Technologies using entirely different concepts may prove to be winners. It is clearly too early to prejudge the concepts that will successfully drive costs down through the end of the century.

Fallacies of Applying the Learning Curve

Forecasting cost reductions in photovoltaics by analogy to the historical learning curve for transistors falters on several grounds. First, each point on the learning curve describes the amount actually sold at prices greater than or equal to the indicated costs. At each cost on the learning curve there was a demand for the product. But there may not be a demand for photovoltaics at every point on the curve.

Photovoltaics are currently competitive in small-scale, remote applications. These include repeating radio signals, collecting meteorological data in mountains, cathodic protection of deep oil wells from corrosion, and illuminating signal buoys. In these applications the costs of conventionally generated electricity may exceed one dollar/kwh. Photovoltaics will have to generate electricity for less than 10 cents/kwh to compete in mass markets. The market in the range between one dollar and ten cents appears to be very thin. Indeed, to attain the ten-cent cost level, the photovoltaic industry may have to make great leaps in investment and production.

Second, much of the progress of the semiconductor industry emerged from miniaturization. The performance of a given volume of material has been increased by many times by condensing ever more complex circuits and shrinking the volume of a given memory. The photovoltaic industry cannot follow this route directly. While the thickness of photovoltaic cells can be reduced, the conversion efficiency of a given area of silicon cells cannot more than double.

The Delphi Method

The Delphi Method seeks a consensual judgment of experts on the likelihood of events that will support a given technological development. Experts are individually asked to identify the most likely date for a particular event. Then they are shown the collective judgment of the panel and asked to reassess their position. After several iterations, a fairly tight consensus usually emerges. This method is more suited to the identification of breakthroughs or discontinuities than to the extrapolation of the learning curve.

The Solar Energy Research Institute ran a workshop in 1978 in which eight industry experts were asked to estimate the year in which thirty events, critical to the success of photovoltaics, were 25 percent, 50 percent, and 75 percent probable. These events included estimates of the cost of alternative fuels, federal tax incentives, and technological developments. Estimates were made under then current policy conditions as a baseline, under conditions of increased federal demand-pull initiatives, and under a doubling of federal research and development funding (Costello et al., 1978). Many of the panel's forecasts appear to be far from the mark.

The panel thought that under baseline conditions it was 50 percent probable that the $2,800/kw cost target for silicon arrays (translated into 1980 dollars) would be reached by 1983 and the $700/kw target reached by 1989. Doubled R&D expenditures and an acceleration of government purchases would accelerate the attainment of these targets by one year each. In 1983 arrays were available for $4,200/kw (in 1980 dollars), far above the 1983 forecast (PI, May 1983).

The panel thought it 50 percent probable that a twenty-year system durability and a 15 percent cell efficiency could be routinely achieved by 1986. Low-cost solar-grade polysilicon ($10/kg) was thought to be achievable by 1984. Low costs ($550/kw) for the balance of system, including a cheap ($20/kwh) battery was thought to be achievable by 1987.

The panel thought it 50 percent probable that by 1982 the commercial market would sustain an annual demand for 100 MW at $2,800 to $4,200/kw (translated into 1980 dollars). Whether the market would be so large at these prices is difficult to determine. The actual world market in 1982 appears closer to the 5-MW to 10-MW range.

The panel expected annual growth in demand to fall to 50 percent in the period 1982–86 as the market grew from 100 to 500 MW. Beyond 1986, growth rates were expected in the 20 to 50 percent range. At these low growth rates, even a high rate of technical progess (.5) would be insufficient to generate 40 percent annual cost reductions. Any further cost reductions would have to come either from static economies of scale or a switch to a different technology.

Major cost reductions resulting from economies of scale are not out of the question as annual output grows from 100 to 500 MW. As an analogy, Texas Instruments reports that each time the volume of its semiconductor production doubles, unit costs drop 30 percent, or to 70 percent of the previous costs (DTH 26 Dec. 1979). A quintupling of annual output reduces unit costs to about 40 percent of their previous level. If similar economies of scale apply to photovoltaics, a quintupling of output alone can reduce costs from $9,000/kw to $3,600/kw. Whether demand will permit such an expansion of output is another question.

The Delphi panel underestimated the rapidity with which political events would favor solar energy. The panel forecasted the legislation of 25 percent investment tax credits, the rise of oil prices to $30/bbl. (in 1980 dollars), the deregulation of some energy prices, and a major nuclear accident by 1981; a second oil embargo by 1984; and the favorable action by utilities toward solar congenerators by 1986.

Nearly all of these events had come to pass by the time the panel's report was published later in 1978 with the passage of the various National Energy Act bills. The Energy Tax Act instituted 20 to 30 percent tax credits for investments in alternative energy devices, a credit raised to 40 percent in 1980. The Natural Gas Policy Act initiated the deregulation of natural gas. The Crude Oil Windfall Profits Tax Act of 1980 began the deregulation of oil prices, which is now completed. The Public Utility Regulatory Policies Act of 1978 mandated the acceptance of sellback by utilities. Some utilities and public utility commissions have already published fairly favorable rates for sellback (SE, Feb. 1980). World oil prices hit $30/bbl. in 1979.

Even though the external environment for solar photovoltaics has been far more favorable than expected by the experts, cost reductions and market growth have been far below expectation. The errors of these experts indicate just how difficult forecasting new energy technologies is.

Microforecasts of Technologies on the Drawing Boards

An alternative to forecasting the performance or cost of a photovoltaic system is to explore the current technological possibilities for its components. Since it takes time to bring a drawing-board concept to commercial application, the prospective technologies of the late 1980s are probably on the drawing boards or in the prototype stage today. At various stages of development are several competing technologies that embody different engineering concepts and semiconductor materials. As the most mature technology, the flat-plate silicon-cell collector will most probably dominate for the rest of the decade. Any cost scenario can be derailed by surprises from private industrial laboratories in the United States, Europe, or Japan.

The five basic processes in the manufacture of flat-plate photovoltaic systems are: (1) preparing the raw material or feedstock for the cells; (2) shaping the material; (3) fabricating the cell; (4) assembling the multicell module; and (5) integrating the module with the control system. In 1980 the cost of a photovoltaic system was about $15,000 per kilowatt. About $9,000 of this cost was for the arrays and the remainder for the control system (Javetski, 1979; Gay, 1980; Russell, 1981). Of the array costs, about $2,000 is due to raw materials (Gay, 1980) and the remainder to cell and module fabrication.

Manufacturing the Raw Material

As indicated by the research director of a major manufacturer of solar cells (Gay, 1980), "The major criteria for selecting solar cell material are that it be a semiconductor, be able to form a junction, have a good absorption of the sun's spectrum, be able to efficiently convert sunlight to electricity, be stable and low cost."

The raw material most commonly used for the fabrication of photovoltaic cells is silicon, whose properties are well understood. The second most abundant element on the earth's surface, silicon is ubiquitous in the form of sand and abundant as quartz, a purified crystalline compound. But despite silicon's abundance, solar-grade silicon is extremely expensive.

Current manufacturing processes are energy-intensive. The quartz (silicon dioxide) is reduced to polycrystalline silicon (polysilicon) in a high-temperature furnace. One engineer (Kelly, 1978, p. 152) notes that "approximately 7000 kwh of energy is required to manufacture the silicon for a one kw array. This means that the device must operate in an average climate for 4 years before it produces as much energy as will be consumed in manufacturing the component silicon."

By 1981 this energy input has been reduced by a factor of three by reducing the thickness of the cells and the waste in slicing. Several processes may be able to reduce the energy costs of solar-grade silicon by an additional factor of 10 to 15, and the economic expenditure from $65/kg to the target cost of $14/kg. Small commercial plants applying two different processes are currently being built. Achieving the cost target would require the construction of a plant producing silicon for 23 MW per year, which is more than four times the entire world demand in 1981 (SE, July 1981, p. 16).

Cells based on rarer materials like gallium arsenide, cadmium sulfide, and copper sulfide are under development. In a joint venture between a glass manufacturer and an oil company, $18 million in private funds has been invested in a prototype plant for producing cadmium sulfide cells. These less expensive materials transform a smaller proportion of photons into electricity. Although they produce only one-quarter as much electricity per square meter as silicon cells, cadmium sulfide cells currently produce electricity more cheaply. The amount of energy used to produce the cells can be regained in less than a year of operation in sunny regions. Their greater area requirements recommend them for remote applications, such as irrigation and rural electrification, where there is considerable near-term market potential throughout the world (SE, September 1980). Engineering designs indicate that costs of $500/kw are achievable with thin-film cadmium sulfide cells—if annual output were 1,000 MW, or 200 times world demand in 1981 (Blythe, 1981). The large area requirements would rule out the use of cadmium sulfide cells in densely packed residential areas, where the larger long-term markets may lie.

Shaping the Material

The most common technique of transforming semiconductor material into cells is the creation of large single-crystal sheets suitable for cell fabrication. This transformation can be achieved by several methods, the most commonly used being the Czochralski process. The purified polysilicon is melted in a receptacle containing a minute quantity of a dopant. The atoms of this dopant have one more or one less electron than the silicon atoms, and they are essential for harnessing the photovoltaic effect. A small seed crystal is immersed in the molten liquid and is withdrawn as a square or cylindrical crystal.

The crystals are then cut into thin wafers, with about 40 percent lost as sawdust. Multiblade slicers may reduce this cutting loss to about 33 percent in the future. Unfortunately, the waste is too impure to be remelted. An important approach to reducing raw-materials costs is the production of thinner cells. Tenfold reductions in thickness of silicon cells appear to be feasible.

While the Czochralski process is a mature method, new facilities promise to reduce costs of production (Matlock, 1979). In addition, other processes for producing low-cost ingots are under development.

An entirely different concept involves growing the crystalline silicon directly in ribbons or thin films. In one process, a cheap ceramic is dipped into molten silicon, yielding cells with only 6 percent efficiency. Another, called the dendritic web, yields cells with 16 percent efficiency (Javetski, 1979). These and the ribbon-to-ribbon and edge-defined film-fed growth processes are currently too expensive and experimental to be counted on in the 1980s (Kran, 1978; Mackintosh et al., 1978).

Instead of growing silicon in thick crystals, another approach is to use silicon in its amorphous state. Like cadmium sulfide, amorphous silicon can be sprayed directly onto a supporting material. Amorphous cells are thinner, thereby conserving materials, and their efficiency approaches 10 percent (Adler, 1980).

Fabricating the Cells and Modules

The manufacture of flat silicon cells involves a large number of operations, many of which are now performed manually. The wafer surface is polished and textured to reduce reflectivity. One surface is doped with a second element with complementary electrical properties. Electrical grids are etched on the surface, which is then plated, solder-dipped, and tested. The individual cells must be connected, assembled in a frame, and encapsulated.

Cell fabrication and module assembly are currently low-volume handicraft activities. The application of mechanized mass-production techniques could reduce 1980 manufacturing costs by a factor of about ten if output were about 100 MW of modules per year. This level of factory output was more than ten times 1980 world demand. While investments in such large-scale production seems to be risky at this time, smaller-scale mechanization is now underway in a joint project of two utilities and Westinghouse (SE, September 1980). Banking on growth in world demand, the subsidiary of an oil company has been operating a mechanized factory capable of 10 MW of annual production (PI, May 1983).

Balance-of-System Costs

The cost of photovoltaic power systems includes not only the modules but also peripheral hardware and services. The hardware includes support structures, power conditioning, and storage (if any). The services include the distributor's markup, site preparation, installation, and possibly an architect's fees.

The support structures for large-scale systems embody mature off-the-shelf technology that is unlikely to enjoy much technological progress. Site preparation and installation involve common construction activities like grading and wiring. Great labor-saving innovations in these activities are unlikely. Although installation may progress down the learning curve, costs are likely to bottom out rather quickly. If photovoltaic arrays can replace conventional parts of a structure, such as roofing tiles, then much of the additional costs of the systems are defrayed.

Power-conditioning equipment includes an inverter, which converts direct current (DC) output of the arrays into common household alternating current (AC). Power-conditioning costs of approximately $500/kw in the late 1970s might be brought down to $150 to 200/kw by technologies on the drawing board (Javetski, 1979). Speculative as these estimates are, the order of magnitude for potential reduction in these costs appears to be far less than for array costs.

Storage technologies show modest potential for improvement by the late 1980s. Lead-acid storage batteries are currently mass-produced for automobiles, and few opportunities for cost reductions through the learning curve or economies of scale appear here. New forms of rechargeable batteries are under development in the automobile industry. As a spillover of electric-car development, advanced batteries at half the price and twice the capacity per pound may be available by the mid-1980s (Stirewalt, 1981). Flywheel technologies may provide an alternative mode of storage as well as current inversion, but recent developments have not been promising (Eldridge, 1980, Table 7).

The photovoltaic fuel-cell system may offer the brightest hope for energy storage. In this technology, cells are replaced by small silicon spheres immersed in an acid that conducts electricity. Electricity generated by the photovoltaic effect liberates hydrogen, which can be stored in a metallic compound. The hydrogen can later be used to generate electricity in a "fuel cell" (SE, July 1980, p. 29).

The institution of time-of-day pricing, as encouraged by the 1978 Public Utility Regulatory Policies Act, should provide a great fillip to the development of energy storage systems. Even in the absence of solar-energy generators, customers would benefit from any devices that permitted the purchase of off-peak electricity for use or even sellback during peak periods. While thermal storage devices have been diffused in Western Europe as a result of time-of-day pricing, cost-effective electric storage devices have not yet been developed (Asbury and Kouvalis, 1976; Everitt, 1981). Any developments in electric storage stimulated by time-of-day pricing would redound to the benefit of solar electric systems.

The Behavior of Potential Manufacturers

A résumé of technologies on the drawing boards indicates that a significant reduction in silicon-array costs is technically achievable during the 1980–86 period (Bickler et al., 1978). Such progress does not unfold automatically from trends, as does a technological imperative. Rather, technology develops as a result of decisions of producers in an industry to invest in research and development in anticipation of profitable sales. If commercial prospects seem bright, an entire supply organization will arise to promote photovoltaics: manufacturers of raw materials, semi-finished inputs, and finished systems, wholesalers, retailers, and installers and financial intermediaries. In other words, firms will invest in bringing down the cost of photovoltaic systems if there is a reasonable prospect that these systems will be profitable at the achieved costs.

The Governmental Track Record

The bulk of basic and applied research on solar photovoltaics has been funded by governments around the world, and much of the market is subsidized as well. Less than 3 MW, two-thirds of world shipments of photovoltaic arrays, is aimed at unsubsidized markets (Murray, 1981). In the United States the federal government has contracted much of the development activity to commercial enterprises.

In the whole experience of federally funded research and development, the National Photovoltaic Program (1976–81) was unique in setting both performance and cost targets. The major federal research and development efforts since the 1940s revolved around weaponry and its spinoffs, such as the space program and the commercialization of atomic energy. Little in these programs suggests that publicly directed research and development programs can meet cost targets.

Carried out almost entirely in private enterprises, weapons-related research and development has aimed for performance specifications, with only modest attention to cost. Examples are airplanes that can attain a certain speed over a certain range, or a manned rocket that can reach the moon and return. Although initial bids on contracts by competitors have a price tag, winning bidders have not always been those who promise the product at the lowest price. More important, delivered prices are almost always more than 60 percent above, and sometimes as much as 400 percent above, initial estimates (Peck and Scherer, 1962; Mansfield, 1971, p. 58). Delivery dates are commonly far later than promised. Errors in estimates have tended to be larger for the more innovative products. Given the great technological uncertainties involved in the development of new weapons systems, unanticipated cost increases, changes in lot size, and the ability of contractors to pass costs along to the buyer, these variances are not surprising. Indeed, the military procurement experience induces skepticism that contractors can estimate the costs and delivery dates of new technologies with much accuracy.

The bulk of federal funding for research and development in the energy area has been channeled into the commercialization of atomic energy. The cost target of making nuclear electricity "too cheap to meter" has never been close to achievement.

The Product-Development Environment

In the innovation process, basic research and the design prototypes are relatively inexpensive, absorbing about 5 to 10 and 10 to 20 percent of total product-development costs, respectively. The costs of tooling for manufacture, start-up, and marketing are higher, absorbing 40 to 60 percent, 5 to 15 percent, and 10 to 25 percent of total product-development costs, respectively. As Mansfield (1971, p. 59) notes:

> By the time a project reaches the development stage, much of the uncertainty regarding its technical feasibility has been reduced, but there usually is considerable uncertainty regarding the cost of development, time to completion, and utility of outcome. . . . There is a long road from preliminary sketches showing schematically how an invention should work, to blueprints and specifications for the construction of production facilities. . . . Frequently, the pilot plant is studied before large-scale production is attempted.

The closer that research moves to the commercialization phase, the larger is the relative private commitment, in order to protect proprietary interests. By 1981 the private sum of funds devoted to PV development had come to equal the sum of public funds (DTH, 1 September 1981).

While the demand for energy-saving investments is easily understood, the behavior of manufacturers on the supply side is harder to predict because of the great risks they face. Potential manufacturers have several options. First, they may refrain from entering a market because of perceived lack of profitability. Two large manufacturers of solar hot-water systems (Exxon and Olin) have recently withdrawn from the market for this reason (SE, July 1981). Second, potential manufacturers can invest in research and development. This strategy commits relatively few resources in a technologically volatile environment and opens the

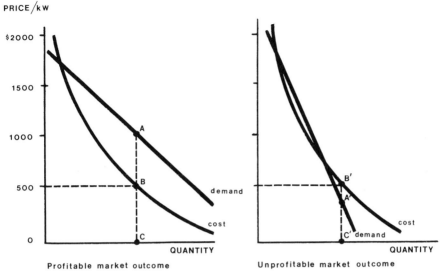

Figure 3.4 Economies of scale and market demand.

possibility for undercutting the competition several years down the line. Most potential photovoltaic manufacturers have chosen this strategy. Third, manufacturers may choose to operate at the handicrafts level and progress slowly down the learning curve. Fourth, a large capital commitment can be made to an existing production technology. This strategy permits the realization of economies of scale and the capture of name identification and a market share. The manufacturer, however, is subjected to the risks that the market cannot support the planned scale of production and that an even newer production technology will make the large investment obsolete (*Figure 3.4*). Only a few companies have made such commitments to photovoltaics, and most of them have matching government funds.

The outcome of these decisions can be placed in the framework of a "payoff matrix" (*Table 3.1*). The producer's options are indicated in the rows and future market environments in the columns. The number in a cell represents the payoff from a given decision in a given market environment. The value of these numbers is arbitrary but illustrative. If the potential manufacturer does not enter the market, no gains or losses ensue. If minor R&D expenditures are incurred, the firm may choose not to exploit a small market later. On the other hand, if the market is large, the firm may be in a good position for its exploitation. If the firm maintains a handicraft level of production, it will make small profits in

Table 3.1 Payoff Matrix for Producer: Industrial Strategy vs. Size of Market

Industrial Strategy	Market Size	
	small	large
Do nothing	0	0
R&D investment only	−2	25
Handicrafts production	10	?
Mass production	−100	500

small markets but will have no experience with the large-scale technologies that become competitive in the larger markets. Finally, if the firm chooses the risky strategy of committing funds to a large-scale plant, it will suffer losses if the market proves to be small. However, it may enjoy great gains if the market proves to be large. How a particular company acts depends not only upon its assessment of the likelihood of various market conditions but also on its assessment of what consumers and competitors are likely to do.

The Paradox of Technological Change

The very prospect of rapid technological change may inhibit the realization of innovative investments. In an environment of rapid technological change, there are considerable advantages to delaying decisions on the part of both producers and consumers. By waiting, both parties can acquire more information about the performance of the innovative technology. Producers can acquire more information about market demand and avoid being locked into a production process or product that may become obsolete overnight. By waiting, the consumer can hope to purchase a superior product at a lower cost. These advantages are obviously greater for more durable products. For products of low durability and cost, like flashlight batteries, there may be little advantage in waiting for something better, but for a commodity with a twenty-year lifetime, like a photovoltaic system, there may be great advantages to waiting. These tendencies may delay investment in full-scale facilities on the part of prospective producers.

This interaction between producer and consumer can be illustrated by considering an additional payoff matrix (*Table 3.2*). Cost expectations are indicated in the rows and consumer actions in the columns. If the consumer expects the costs of the new energy technology to remain stable, it is better to save conventional energy now rather than later. On

Table 3.2 Payoff Matrix for Consumer: Cost Expectations vs. Timing of Investment

	Timing of Investment	
Cost Expectations	now	wait
Decreasing	2	10
Stable	5	1

the other hand, if costs are expected to fall, it is better to wait. The problem for the manufacturer is that if he expects consumers to hold back, he will not invest in production facilities and costs will not in fact come down.

The rapidity of technological change in the photovoltaics industry, the uncertainties about the "winning" technology, and the likelihood of a rapid shakedown and concentration of production in a few large-scale firms can be another serious impediment to the actual commitment of investment funds to production facilities. There are few advantages and many disadvantages to a producer's plunging into immediate full-scale production. A pioneering firm may stake out a high market share by virtue of its primacy. Traditional market shares tend to be more important for commodities of low durability, where repeat purchases are routine. Permanent market shares may be less common in innovative industries that produce durables, like electronics. A pioneering company that made a major investment might find its technology eclipsed by a follower who invested several years later. The advantages of being second in line would appear to be greater than the advantages of being first. Unless a company believes its technology has across-the-board superiority, it will hold back, and so will other companies.[3]

A particular company's decision to enter the market first depends upon its expectations of its rivals' capabilities. In Table 3.3, its options are indicated as rows in the payoff matrix and its expectations as columns. If the firm enters first and the rival's technology is no better, then primacy confers its advantages. But if the rival's technology is initially superior, then the fool ventures first. If the firm chooses to wait and the rival's technology is no better, then the rival wins the advantage of name recognition. If the rival's technology is initially superior, then it pays to wait and outflank the rival with a new technology later.

This paradox has always existed in an innovative environment. As Joseph Schumpeter noted, the "herdlike swarm" of followers is an essential part of the process of "creative destruction," in which the leader often perishes.

Table 3.3   Payoff Matrix for Producer: Timing of Entry vs. Assessment of Rival's Technology

Timing of Entry	Assessment of Rival Technology	
	no better	superior
Enter first	50	−100
Enter second	10	100*

*Assumes that firm can surpass rival's initially superior technology in time.

The paradox of rapid technological change is that the technical feasibility of achieving 1986 cost targets is no guarantee that the production facilities for realizing these targets will be built. If manufacturers perceive the potential market size to be small, if they perceive that consumers expect continued technological progress, or if they perceive great risks in making a premature commitment to a rapidly advancing technology, then the expectations of technological change may prove to be self-defeating.

Characteristics of the Solar Industry

Membership in the photovoltaics industry is not easy to define. Involvement in the industry ranges from undertaking applied research, to conceptual design of products, to planning pilot plants, to full-scale production. Firms that are not participating in any of these activities today, particularly energy or electronics firms, might easily enter the industry tomorrow by acquiring an existing company or constructing new facilities.

As a classic infant industry, the solar photovoltaics industry is comprised of participants who vary in size, degree of vertical integration, and financial capability. The industry includes giants in the chemical industry, subsidiaries of oil companies, and independent firms. Some have been traditionally involved in semiconductors, while others are new to the field. Most firms specialize in a single process, while a few are vertically integrated from raw-materials fabrication to module assembly.

The proliferation of firms is common in the infancy of industries, from automobiles to calculators. To the extent that economies of scale are large relative to the prospective market, a shakedown is inevitable. Promising firms may expand or be acquired as less promising firms merge or perish. Nearly all experts foresee a rapid concentration of the industry, with a single company supplying 40 percent of the world market by 1985 (Costello, 1978). The fact that three firms control 80 percent of the world market may be indicative (SEC, March 1982, pp. 16–18).

The linkage between photovoltaics producers and large companies in the oil and electronics business may provide a resolution to the paradox of self-defeating technological change. Accustomed to undertaking large-scale, risky investments with long payback periods, these companies have operating procedures and financial resources supportive of start-up and expansion stages of innovative industries. They have the financing resources to practice "forward pricing." This refers to pricing a new product at a cost achievable in the future after the firm has progressed down the learning curve and has expanded the scale of production. If a firm practices forward pricing, then its costs will exceed revenues for a short time. Indeed, it appears that several solar subsidiaries of large corporations are engaging in this practice, which has been called "predatory" by well-meaning but self-styled spokesmen for solar energy (SEC, June 1982, p. 4). Forward pricing can only expand the current demand for solar photovoltaics by reducing the consumer's advantage of waiting. This practice provides one method of defeating the paradox of self-defeating technological change.

Notes

1. While there may be temporary aberrations, prices are proportionate to costs, including normal profits in fabrication, marketing, and installation. Innovators who monopolize a new technology have the power to earn abnormally high unit profits, but they may choose instead to price below costs (and suffer abnormal losses) in order to create a market. The large number of international competitors in the transistor and photovoltaic industries ensures that prices stay in line with costs.

2. Until the early 1970s, when nearly all solar cells were produced for the U.S. Government by two companies, total output was easy to gauge. The subsequent proliferation of producers and the growth of the commercial markets make accounting for production somewhat difficult. Thus recent estimates of cumulative world production may be inaccurate (Murray, 1981).

3. Another offshoot of semiconductor technology, the personal computer market, may illustrate the point. By 1979 the existence of a mass market for 48K personal computers priced below $3,000 was proven by the Apple and to a lesser extent by the Radio Shack Model II. Giant IBM was a latecomer to the market, but it made a crash effort to set superior performance-to-price standards. IBM is now the leader in market share, while the initial leaders have been losing share (BW, 11 July 1983).

4
The Emerging Photovoltaic Markets

More than just another investment, like an addition to a house or the replacement of a machine, a solar photovoltaic system is an innovation, a new way of performing an old function. While businesses and consumers are affected by somewhat different factors in deciding to adopt an innovation, the major ones are fairly similar (Roessner et al., 1979). First is the profitability or net benefit of the innovation over the conventional way of doing things. In the case of solar-energy systems, the net savings in fuel or electricity costs is the most important factor for utilities, other businesses, or consumers.

Second, the greater the riskiness of an innovation, the greater the expected return that is necessary to induce adoption. Solar-energy systems entail greater technological and institutional risks than conventional energy systems, but they provide insurance against increases in conventional energy prices. This insurance characteristic suggests that solar-energy systems may be desirable even if they cost slightly more than the expected value of the conventional energy saved.

Third, innovations that entail expensive and irreversible commitments must offer a high ratio of benefits to cost to be attractive. Unlike an ordinary perishable product, which can be tried and discarded if unsatisfactory, solar-energy systems have an expected life of about twenty years. The magnitude of the financial commitment and durability of solar-energy systems offsets to some extent their insurance advantage.

Less important in the corporate context, personal characteristics influence the likelihood of consumers' adopting innovations. Better-educated consumers are better able to evaluate evidence and more willing to take risks than less educated consumers. Wealthier individuals are able to resist the social pressures to conform to conventional ways and even to take pride in being in the vanguard.

Penetration of decentralized photovoltaic systems to mass markets may require forms of ownership that reduce the financial commitment and risk to the ultimate user. These considerations suggest that leasing, patterned after utility-owned telephones, may provide the method of penetrating the middle- and lower-income residential markets. Thus the adoption of solar-energy systems should be analyzed from the viewpoint of both high-income individuals and utilities that own and lease these systems.

This chapter examines the returns on two applications of photovoltaics in several regions of the United States. Irrigation is an example of a remote application in which actual energy use can be accommodated to the availability of sunlight. Residential photovoltaics exemplify a utility-interactive application, where patterns of energy use are largely independent of patterns of insolation. Regional differences in the initial viability of solar photovoltaics are identified as well as the optimum year of investment under various scenarios.

Agricultural Photovoltaics

In the United States more than 35 million acres, or 10 percent of all farmland, is irrigated. Practiced almost exclusively west of the Mississippi, irrigation generally takes place on large farms (160 acres or more) with large pumps (100 to 300/kw peak power). Irrigated agriculture produces 20 percent of American farm output and almost all the output in states like Arizona and New Mexico (Dvoskin and Heady, 1976; Sloggett, 1977). Only .5 percent of the nation's energy budget is expended on irrigation, amounting to $17 per acre in the early 1970s. What these averages belie is the critical importance of energy costs for irrigated agriculture in particular locations.

Irrigation may be a particularly apt application of solar-energy photovoltaics for four reasons. First, the need for pumping is coincident with the period of maximum insolation. This is because evapotranspirative losses of water are proportional to insolation. Second, in advanced countries the costs of extending electrical wires to widely spaced irrigation pumps are relatively high. In developing countries, where the electrical grid may be remote from villages, the cost of transporting fuels for internal-combustion generators is correspondingly high. Third, the irrigation loads generally peak in the months when electric utility loads peak. Consequently, utilities often impose stiff peaking fees upon irrigators. For this reason, 80 percent of the energy used in pumping in the United States is generated by on-site internal-combustion engines (Sloggett, 1977). The thermodynamic efficiency of these engines is less than 25 percent, compared to about 33 percent in central electric plants. Fourth, photovoltaic generators are amenable to dispersion because economies of scale in their use are small.

The Critical Role of Irrigation Costs in Arid Regions

The viability of irrigation in arid regions is highly dependent on energy costs. The amount of energy consumed in irrigation depends upon the length of the growing season, the seasonal distribution of rainfall, the height that surface and ground water must be lifted, and the price or availability of water. In descending order, Texas, Nebraska, Arizona, New Mexico, and California consume 70 percent of all irrigation energy used in American agriculture (Sloggett, 1977).

The time when conventionally powered irrigation becomes unviable can be computed through simple assumptions about land values and rent. Agricultural-land rent equals the difference between operating costs and operating revenues, including returns on reproducible capital and entrepreneurship. First, assume that real energy costs increase at the compound rate r, but that other prices and crop yields exhibit no long-term trend. Second, assume that agricultural rents are approximately 3 percent of land values, as has been the case historically (Ferraro, 1969; Castle and Hoch, 1982). Third, ignore the falling level of the aquifer, which increases the expenditure of physical energy required to lift water.[1] Irrigation becomes unviable when the increment in fuel costs above the current levels F_0 equals the current difference between irrigated and unirrigated land rents. I is the initial value of irrigated land, U is the initial value of unirrigated land in the region, and a is .03:

$$a (I - U) = (F_0 e^{rt} - F_0), \text{ or}$$
$$rt = \ln (F_0 + aI - aU) - \ln F_0$$

For example, in 1975 the value of irrigated land in western Nebraska was $2,000/acre; the value of unirrigated land was $900/acre. Diesel fuel

Table 4.1 Years in Future When Rising Fuel Costs Make Irrigation Unviable, Base Year 1975

	Producing* Region	Assumed Real Annual Growth Rate of Fuel Costs		
		0%	2%	4%
Nebraska	55	∞	87	44
Texas	79	∞	33	17
Arizona	87	∞	18	9
California	101	∞	89	45

*Defined by Water Resources Council; see Dvoskin and Heady (1976).

costs were $7/acre.[2] If fuel costs continued to rise 4 percent per year, then irrigated agriculture becomes unviable forty-four years later, in the year 2019. In some parts of the West, the aquifer probably will have been depleted well before then.[3]

Similar calculations are performed for several western "producing regions," defined by the Federal Water Resources Council. These suggest that Doomsday may be only a generation away in Texas and Arizona (*Table 4.1*). If fuel costs escalate at 4 percent, irrigation becomes unviable in southern Arizona in 1984 and in west Texas in 1992. Under conventional energy systems, the practice of irrigating arid lands may wither away, while agriculture shifts to more humid regions.

The Profitability of Solar Photovoltaics

The profitability of substituting photovoltaics for conventional fossil-fuel generators is considered from the public-policy and farmer's point of view in these same four producing regions. To facilitate comparisons, the analysis is taken on a per-acre basis.

Photovoltaic array costs are assumed to fall at the historical annual rate of 26 percent until the year 2000. The balance of the system includes a small battery to smooth the load on the pump during the daylight hours, an inverter to operate an AC pump,[4] electronic controls, and supporting structures. Achievable with current technologies, the balance of the system cost of $1,250/kw is not expected to fall during the period analyzed. Annual maintenance costs are assumed to be 1 percent of the original system cost.

The hypothetical pumping system operates on a stand-alone basis, with neither sales nor purchases of electricity from the utility. There-

68 Solar and Wind Energy

Table 4.2 Irrigation Demand and Insolation

	Producing Region			
	Nebraska	Texas	Arizona	California
Irrigation demand/acre				
1. Thous. Btu to apply 1 acre-foot	1.330	4.909	7.096	1.133
2. Acre-feet applied/acre	1.83	1.50	5.50	3.17
3. kwh for irrigation/acre	164	495	2621	241
4. kwh for crop drying/acre	122	0	0	0
Insolation				
5. Pumping months	July–August	April–September	March–October	March–October
6. Annual kwh/kw	2040	2818	2639	2546
7. On-season kwh/kw	427	1555	1940	1908
8. kw peak required	.384	.318	1.35	.126

Source: R. W. Matlin and M. T. Katzman, "The Economics of Adopting Solar Photovoltaic Energy Systems in Irrigation," MIT Lincoln Laboratory, COO/4094-2, December 1977 (Tables 3 and 6).

fore, the photovoltaic systems are sized to meet all of these energy requirements for the irrigation season. Energy requirements per acre depend upon the relative availability of surface and ground water, the depth of pumping, pump efficiency, and volume of water customarily applied. Energy requirements are low in Nebraska, where the irrigation season is short and the water table is close to the surface. Arizona's immense energy requirements result from the great depth of the water table, the length of the irrigation season, and natural aridity (*Table 4.2*).

Insolation in the Southwestern producing regions is 25 to 40 percent higher than in Nebraska. Because of the shortness of Nebraska's irrigation season, on-season insolation is 300 percent higher in the Southwest. In Nebraska, 122 kwh/acre is devoted to crop drying off-season, whereas none is allocated in the Southwestern regions.

Liquid fuels (diesel oil and petroleum gas) and natural gas are valued at about $7.50/MBtu in 1980 dollars. The on-site generator's 23 percent conversion efficiency implies an energy cost of 11 cents/kwh. The capital and maintenance cost of the generator is taken at 1 cent/kwh, an extremely low figure. The value of energy used in pumping, then, is assigned a value of 12 cents/kwh. Thermal energy for crop drying is assigned a value of $7.50/MBtu or 7.5 cents/kwh. Off-season, surplus

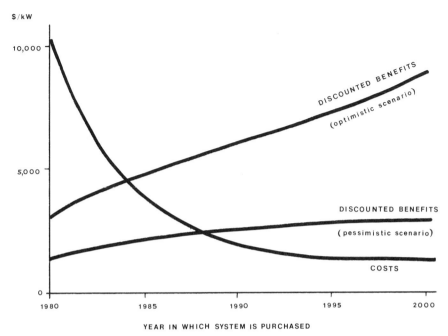

Figure 4.1 Discounted benefits and costs per kilowatt, Arizona agriculture.

solar electricity may be applied to residential uses or to other farm uses, such as the heating of animal pens, but this off-season energy is assigned no value in the present analysis.

The profitability of photovoltaics is examined in each producing region for investments undertaken alternatively in each year during the period 1980–2000. The public-policy analysis is undertaken with and without a 25 percent national-security surcharge on conventional fuels. The alternative fuel-escalation rates are 2, 3, and 4 percent. From the public-policy perspective, the alternative discount rates are 5 and 10 percent. From the family farmer's perspective, the aftertax discount rates are 5 and 8 percent. The farmer enjoys a 25 percent combined (ordinary plus energy) tax credit and depreciates the system in five years.

The time path of social costs and discounted social benefits can be illustrated on a per-kilowatt basis for Arizona (*Figure 4.1*), the most favorable location, and for Nebraska, the least favorable location (*Figure 4.2*). In both locations, total systems costs per kilowatt fall, while net discounted benefits rise through the year 2000. The lower-benefit curve reflects the most pessimistic scenario (10 percent discount/2 percent fuel

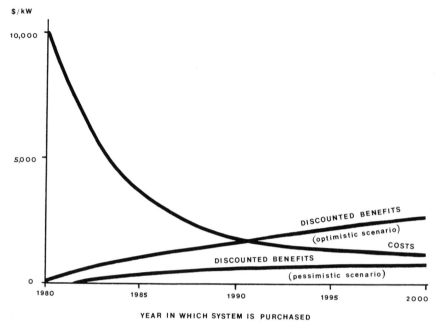

Figure 4.2 Discounted benefits and costs per kilowatt, Nebraska agriculture.

escalation), and the upper-benefit curve reflects the most optimistic scenario (5 percent discount/4 percent fuel escalation). For each benefit scenario, photovoltaic-powered irrigation becomes viable where the lines cross.

Under all scenarios, it is profitable to adopt photovoltaics for irrigation by the late 1980s in all states but Nebraska. Under the most optimistic scenario (5 percent discount/4 percent fuel escalation), photovoltaics become viable in the three southwestern states in the mid-1980s, but after 1990 in Nebraska (*Table 4.3*). Under the most pessimistic scenario (10 percent discount/2 percent fuel escalation), photovoltaics become viable around 1990 in the three southwestern states and well after the year 2000 in Nebraska. The difference in initial viability between the most pessimistic and most optimistic scenarios is only about five years in the southwestern states, but it is more than ten years in Nebraska.

Under all scenarios, photovoltaics become competitive with conventional energy before fuel costs render irrigated agriculture unviable. Like a paladin, photovoltaics rescue irrigated agriculture from an imminent demise in Arizona. The limiting factor becomes the depletion of the aquifer, not energy costs.

If a 25 percent national-security surcharge is levied on conventional fuels, then photovoltaics become profitable three years earlier in Nebraska but only one year earlier in the Southwest. Compared to the public-policy perspective, the benefits of photovoltaics from the farmer's perspective are reduced somewhat by income taxes, while the costs are reduced proportionately more by tax credits and depreciation allowances. In all four states photovoltaics become profitable to farmers about two or three years before indicated from the public-policy point of view.

The Optimum Investment Year

The initial year of viability is not necessarily the best year in which to purchase a photovoltaic power system. The advantage to waiting is the anticipation of falling systems prices. The optimal year of investment is that in which net present worth (NPW), discounted to the present (1980), is maximum (j is the year of investment and i is the discount rate):

$$\text{maximize over } j \quad NPW_j/(1+i)^j$$

So long as the NPW increases faster than the discount rate, it pays to wait. This point can be illustrated by the most pessimistic scenario for Arizona (*Figure 4.3*). Net present worth rises at a decelerating rate in the period 1980–2000. Discounting this stream to 1980, one obtains a peak in the year 1989, which is the optimal year of investment for that scenario.

In Arizona another constraint must be considered in selecting the optimal year of adoption. Conventional irrigation becomes unviable there in 1993 if fuel prices rise by 2 percent annually. Farmers cannot wait any later than that year either to abandon irrigation or to switch to photovoltaics. If fuel costs rise at 4 percent, they can wait no later than 1984.

Except for Arizona, the optimal year of investment comes six to fifteen years after initial viability. The lag is obviously longer for the lower discount rates and higher-fuel escalation rates, but the differences are small. In other words, the prospects for rapid cost reductions may discourage the appearance of a commercial market in irrigation, even though the photovoltaic system could save the farmer money. That is because he can save even more by waiting.

Residential Photovoltaics

The residential sector consumes about 20 percent of the energy utilized in the United States and even a higher percentage in less developed countries. More than half of American residential energy use is for space

Table 4.3 Years of Demise of Conventional Irrigation, Initial Viability, and Optimality of Photovoltaic Irrigation

	Year of demise	Perspective					
		Public policy				Family farmer's	
		(0% externality)		(25% externality)			
		viable	optimal	viable	optimal	viable	optimal
Nebraska							
r = 2%/i = hi*	2062	2000+	2000+	1998	2000+	1996	2000+
r = 2%/i = lo	2062	1997	2000+	1992	2000+	1992	2000+
r = 4%/i = hi	2019	1995	2000+	1992	2000+	1991	2000+
r = 4%/i = lo	2019	1991	2000+	1989	2000+	1989	2000+
Texas							
r = 2%/i = hi	2008	1990	1996	1989	1994	1987	1993
r = 2%/i = lo	2008	1987	1996	1986	1995	1986	1994
r = 4%/i = hi	1992	1988	1989	1987	1993	1986	1993
r = 4%/i = lo	1992	1986	2000+	1985	2000+	1984	1998
Arizona							
r = 2%/i = hi	1993	1988	1994	1987	1993	1986	1992
r = 2%/i = lo	1993	1986	1995	1985	1994	1984	1993
r = 4%/i = hi	1984	1987	1993	1986	1992	1985	1992
r = 4%/i = lo	1984	1985	1999	1984	1998	1983	1996
California							
r = 2%/i = hi	2063	1988	1993	1987	1993	1986	1992
r = 2%/i = lo	2063	1986	1995	1985	1994	1985	1993
r = 4%/i = hi	2020	1987	1993	1986	1992	1985	1994
r = 4%/i = lo	2020	1985	1999	1984	1998	1983	1997

r = fuel-escalation rate.
i = discount rate; hi = 10% for public policy, 8% for family farmer; lo = 5% for both public policy and family farmer.
Farmer's marginal income tax rate = 30%.

heating, and less than one-sixth for water heating, trailed by refrigeration, air conditioning, and other minor uses (Schurr, 1979, Table 2–3).

In the United States more than half of this residential energy is consumed directly as gas (34 percent) and oil (19 percent) and the rest as electricity (Hirst and Carney, 1977, Table 3). The residential sector may become increasingly electrified as the prices of gas and oil rise relative to coal, the future mainstay of electric utilities. Photovoltaics, then, are a potential competitor to conventional electricity in the entire gamut of residential energy demands.

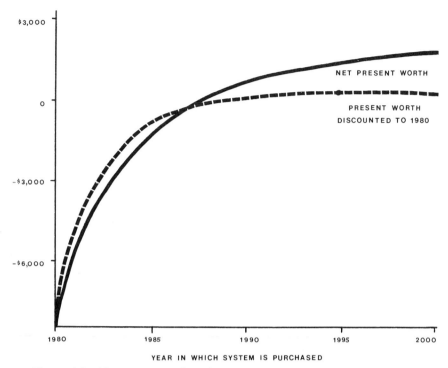

Figure 4.3 Net present worth and present worth discounted back to 1980: Arizona agriculture, pessimistic scenario.

The Technical Options

The performance of a large number of residential solar systems has been examined in Fort Worth, Texas, and in New York City by MIT Lincoln Laboratory (Russell, 1979). The baseline is a well-insulated all-electric home of 1,800 square feet (about 160 square meters). In all systems, most space heating and all cooling loads are served by a high-efficiency heat pump. In the winter the heat pump can draw thermal energy either from the air or from a water-storage tank. An electric-resistance heater provides backup in the winter.

For solar systems the two electric options are an AC utility-interactive system and a DC system with battery storage. Collectors are of three types: all photovoltaic, photovoltaic and thermal collectors laid side by side, and thermal collectors overlaid on the photovoltaic collectors (a hybrid). Thermal collectors serve as a source of energy for domestic hot-water and space heating. They can also be used to preheat coils of the heat pump during the winter. Total collector area varies within the range of 25 to 100 square meters, which is the roof area of an average single-family home. Thermal storage varies from 50 to 200 kg/square meter of thermal collector.

The performance of each configuration was simulated by the TRNSYS computer algorithm of the University of Wisconsin Solar Energy Laboratory. Weather data, including insolation, were obtained from the National Oceanic and Atmospheric Administration's SOLMET weather tapes for "typical meteorological years" in the two sites. Temperature-sensitive space-conditioning loads are calculated by formula, while the remaining loads were estimated from a study by the Bureau of Standards. The output of each simulation is electricity purchased and sold back during the peak, shoulder, and base periods, which were defined by the respective utilities. The electric loads of each solar house is compared to the no-solar baseline house.

The Economic Parameters

The value of electricity savings is determined by the rate structures of the utilities in New York City and Fort Worth. Both utilities offer homeowners experimental rate structures that vary by season and time of day. Rates are lowest when demand on the utilities is low and the most efficient plants, using the cheapest fuels, can be exploited. When demand is at its peak and the least efficient plants are burning the most expensive fuels, rates are highest. Peak rates are also high because they reflect the costs of assuring that sufficient generating capacity is available.

Encouraged by the Public Utility Regulatory Policies Act (PURPA), time-of-day rate structures are likely to become more common as utilities attempt to reduce peak loads and thereby ease their construction requirements. On the supply side, utilities will continue their great transition from gas and oil toward coal as a base- and shoulder-load fuel. In New York State oil will continue to serve as a peaking fuel, while in Texas natural gas will play this role (NERC, 1983). Even if coal-derived synthetics replace gas and oil as a peaking fuel (EPRIJ, March 1981), serving peak loads will continue to be expensive.

Both load management and the conversion to coal generation will change the cost structure of utilities. These are likely to alter the time-of-day rate schedules offered. Because these changes are difficult to anticipate, assume here that current energy-use patterns and rate structures prevail. In Fort Worth the 1980 costs are 15 cents/kwh during the peak, 4 cents/kwh in the shoulder period, and 1.5 cents/kwh in the base period. In New York City the corresponding costs per kilowatt hour are 18 cents, 8 cents, and 3 cents, respectively. The estimated fuel portion of these rates is escalated 2 to 4 percent per year over the life of the solar investment.

What rates will utilities offer for electricity they purchase from customers? Both PURPA and the Electric Power Research Institute, the research and development arm of the electric utilities, encourage the consideration of marginal costs in rate setting. These are the costs of generating one more or one less kilowatt-hour of electricity at a particular time in the load cycle. If electricity is priced according to a marginal-cost principle, the utility costs avoided by reducing demand by one kilowatt-hour should nearly equal the cost avoided by receiving one kllowatt-hour from a customer. In other words, if a peak kilowatt-hour is sold to the consumer for 16 cents, the utility also ought to pay back nearly 16 cents for a kilowatt-hour received. Losses from transmission and distribution back to the grid should reduce the amount received for sellback by about 10 percent. We assume that utilities will pay about 90 percent of what they charge consumers for each additional kilowatt-hour.[5]

The cost of photovoltaic arrays is projected to continue falling at the historical 26 percent annual rate to the year 2000, and the balance-of-system cost is projected to fall 20 percent each year until 1990. Because of their technological maturity, solar-thermal systems are expected to decrease in cost at only 5 percent annually until 1990.

The economic analysis is undertaken for each year from 1980 to 2000 from three points of view: the public-policy maker, the homeowner, and the utility. Public-policy calculations are performed with and without a 25 percent surcharge on fuel for externalities. The homeowner is assumed to fall in the 30 percent marginal tax bracket and to receive a 40 percent credit without limit.[6] A utility subsidiary is in the 46 percent marginal tax bracket and receives the 10 percent investment tax credit plus a 15 percent energy-saving tax credit. The initial year of profitability, the profitability in 1990, and the optimal year of investment are determined.

Public-Policy Results

The initial public-policy analysis excludes external costs. Even under the best scenario for the solar industry (5 percent discount/4 percent fuel escalation), residential photovoltaics become viable from a public-policy point of view well after the 1986 target date of the National Photovoltaic Plan (*Table 4.4*).

Even though solar-energy systems produce a third less output in New York City than in Fort Worth, they become viable earlier in New York City, which has far higher electricity rates. In Fort Worth only the all-photovoltaic systems become viable by the mid-1990s under the most optimistic scenario. Under lower fuel-escalation rates or higher discount rates, no solar-energy systems are viable from a public-policy perspective

76 Solar and Wind Energy

Table 4.4 Initial Year in Which Residential System Becomes Economically Viable, Public-Policy Perspective (excluding external costs)

		New York City				Fort Worth			
Fuel		2%		4%		2%		4%	
	Discount	5%	10%	5%	10%	5%	10%	5%	10%
Therm m^2	Pv m^2								
Hybrid*									
25	25	1990	1995	1989	1991	2000+	2000+	1998	2000+
50	50	1989	1991	1988	1990	2000+	2000+	1996	2000+
75	75	1993	2000+	1990	1996	2000+	2000+	2000+	2000+
100	100	1995	2000+	1991	1997	2000+	2000+	2000+	2000+
All-PV**									
0	25	1990	1994	1990	1992	2000+	2000+	1994	2000+
0	50	1991	1994	1990	1992	2000+	2000+	1994	2000+
0	75	1991	1995	1990	1992	2000+	2000+	1994	2000+
0	100	1991	1995	1990	1992	2000+	2000+	1994	2000+
Side-by-side*									
25	25	1990	1993	1989	1990	2000+	2000+	1996	2000+
50	25	1993	2000+	1990	1993	2000+	2000+	2000+	2000+
75	25	1996	2000+	1990	1998	2000+	2000+	2000+	2000+
50	50	1990	1993	1987	1992	2000+	2000+	1997	2000+
25	50	1989	1992	1988	1990	2000+	2000+	1996	2000+

*Storage of 50 kg./m^2.
**No electric storage.

before the year 2000. In New York City a few systems achieve viability before 1990, but most become viable under a wide range of scenarios by the early 1990s.

The larger photovoltaic systems deliver a larger proportion of their output to the utility. For example, a 50m^2 (5kw) system in New York City may sell back only 15 percent of its output, while a 10kw may sell back over 40 percent. The larger the system, the greater is the dependence of economic viability on the sellback rate.

For larger systems the discounted present worth of the electricity sold back more than compensates for as much as $700 worth of the additional electrical equipment required to prepare the utility for sellback (Kammer, 1979). In both cities the utility-interactive AC systems were far superior to stand-alone DC systems with electric storage.

When a 25 percent premium for external costs is added to the fuel portion of the electricity rates, photovoltaics become viable somewhat earlier. The effect of including this premium is not great, however. The dates in the scenarios of Table 4.4 are advanced only one to four years. In New York City no systems are viable until 1987 under even the most optimistic scenario. In Fort Worth photovoltaics are still not viable until the 1990s.

Results for Homeowners

Because of tax credits and depreciation allowances, systems become viable earlier from the utility perspective than from the public-policy perspective. The 40 percent tax credit has the effect of advancing the viability of systems even more so from the homeowner's viewpoint. Homeowners may find a system viable as much as ten years before it is viable from a public-policy point of view.

From the homeowner's perspective a few all-photovoltaic and side-by-side systems become viable in New York City by 1987, shortly after the target date of the National Photovoltaic Plan. All of the systems become viable by 1990 from this perspective. Which of these viable systems is the most profitable?

A side-by-side system with 25m^2 of thermal collectors and 75m^2 (7.5kw) of photovoltaic collectors yields the highest net present worth by far. Two other side-by-side configurations surpass the best all-photovoltaic system of 100m^2 (10kw). While net present worth increases with the area of the photovoltaic arrays, it appears to decrease with the thermal array area (*Figure 4.4*). This is because larger photovoltaic systems can always sell surplus electricity to the utility, but larger thermal systems cannot easily sell surplus heat.

Under the low fuel-escalation scenario, no system appears to be viable to Fort Worth homeowners until the 1990s. In 1991 the four all-photovoltaic systems become profitable, with the 10-kw system producing the highest net present worth. Small side-by-side systems (25m^2) become viable a few years later.

These calculations of viability are based upon the criterion of net present worth. This figure of merit is expressed arithmetically as:

$$\text{NPW} = \text{BENEFIT} - \text{COST}$$

Another figure of merit is the benefit-cost ratio:

$$\text{B/C} = \text{BENEFIT/COST}$$

The two criteria provide identical indications of whether or not a particular investment is profitable. Whenever the net present worth is greater than zero, the benefit-cost ratio is greater than one.

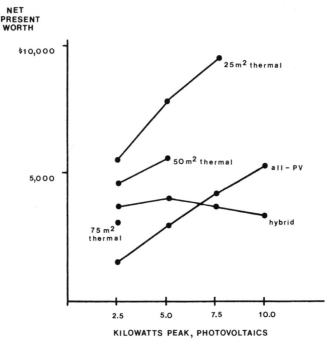

Figure 4.4 Net present worth of best residential solar systems, 1986: New York City, homeowner's perspective.

The magnitude and irreversibility of solar investments may lead homeowners to prefer the configuration that yields the highest benefit-cost ratio rather than the highest net present worth. In choosing among mutually exclusive investments, maximizing the net present worth does not always result in maximizing the benefit-cost ratio.

For example, the three systems with the highest net present worth in New York City in 1990 are side-by-side configurations with 25m^2 of thermal collectors and with 75m^2, 50m^2, and 25m^2 of photovoltaics. At the lower fuel-escalation rate, the benefit-cost ratios of these systems are 2.05, 2.12, and 2.22, respectively. Homeowners averse to large, irreversible commitments might select one of these smaller systems, thus rendering the market for photovoltaics smaller than predicted on the basis of net profitability alone.

Levelization and Utility Ownership

Despite the apparent profitability of residential solar systems in the late 1980s, homeowners may be reluctant to purchase, finance, and maintain them. Instead, utilities may establish subsidiaries to install and maintain these systems on a leased basis. What would utilities have to charge to recoup the costs of the systems?

As an illustration, the optimum system in New York City costs about $13,100 (before tax credits) in 1990. It reduces annual purchases of electricity by 8,700 kwh and permits the homeowner to sell back 4,850 kwh, energy with a present value of about $18,200. The utility might lease the system to the homeowner on one of several bases.

First, a fixed annual fee may be charged, like a level mortgage payment. At a 5 percent discount rate, this annual payment would be about $1,200, quite a bit lower than the levelized value of the energy saved, nearly $1,500.

Second, the utility might charge a fixed price P for each kilowatt-hour used by the homeowner and offer $.9P$ for each kilowatt-hour fed back. The value of P that would permit the utility to recoup systems costs is 9.2 cents/kwh, below the levelized value of the energy saved, 11.4 cents/kwh. This charge is below the expected peak-load and shoulder-load prices in 1990 but above the expected base-load prices.

Third, instead of charging a fixed annual fee or fixed rate per kilowatt-hour, the utility might escalate these fees or rates along with conventional electricity prices. This would provide the homeowner with lower costs in the beginning and higher costs as benefits escalate. The arithmetic of such schemes is elaborated in the Appendix.

The Paradox of Technological Change

Even after a solar-energy system becomes profitable, system costs are expected to continue falling, while benefits will continue to rise with fuel escalation. There may be an advantage to waiting as much as ten years after initial viability before purchasing a photovoltaic system.

What is the optimal time to buy? In our examples, at the lower fuel-escalation rate, the best system in New York City ($25m^2$ thermal/$75m^2$ photovoltaic) becomes viable in 1987. The net present value of the system is $429. If we wait an additional year, the present value becomes $4,045, nearly ten times higher. Each year we wait, the net present worth increases (*Table 4.5*).

Because future values are discounted, it is not obvious that we should wait forever. The value of the system purchased in 1988 is only $3,852 when discounted back to 1987. The value of a system purchased in 1992 is only $8,823 when discounted back to 1987. In other words, it is worth waiting at least until 1992 before purchasing this system. Waiting longer, however, is not economical because the absolute value of the cost reduction is so small that it is offset by the discount factor. By similar

Table 4.5 The Effect of Tax Credits on Net Present Worth of the Optimal Residential System in New York City (2% fuel escalation/5% discount)

	Without tax credits		With tax credits		
Year of Investment t	System Cost	NPW in year of investment	NPW in year of investment	NPW discounted to 1987	Recommended tax credit (%)
1987	26303	−10092	429	429	62
1988	21580	−4780	4045	3852	40
1989	17804	−144	6978	6329	31
1990	14770	3460	9372	8096	16
1991	13631	4957	10409	8564	8
1992	12745	6162	11260	8823	0

Notes: Recommended reduction = $(NPW_T - NPW_{T-t}) \times 1.05^{(T-t)}/COST$ where T = 1992.

calculations in Fort Worth, the best system becomes viable in 1993, but it is worth waiting beyond the year 2000.

If consumers anticipate advantages to waiting, the growth of market demand will be delayed. If producers perceive that consumers are holding back, they may in response postpone investments in production facilities to realize these cost reductions. Indeed, an analogous process occurred in 1977–78, when Congress debated the tax-credit provisions of the National Energy Act. In anticipation of large credits, which could reduce the out-of-pocket cost to homeowners, potential purchasers held back until the passage of the act. There was little growth in the sales of solar-thermal collectors in the eighteen-month interim, in contrast to annual growth in excess of 40 percent in earlier and subsequent periods (SE, June 1980, p. 42).

How Tax Credits Resolve the Paradox

Tax credits can play an important role in resolving the paradox of self-defeating technological change. While it is worth waiting until 1992, the optimal system would be worth purchasing as early as 1990 even without a tax credit. Suppose we eliminated the solar-energy tax credit and tried to design a tax system that would discourage waiting. The gain to buying in 1992 instead of 1991 is only about $250 (discounted for one year). This is less than 8 percent of the 1991 cost. An 8 percent tax credit in

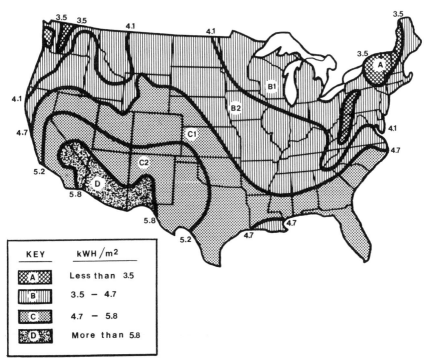

Figure 4.5   Mean daily solar radiation in the United States, kwh/m². Source: after John L. Baldwin, The Climate of the United States *(Washington, D.C.: U.S. Dept. of Commerce, 1973), map 55.*

1991 is sufficient to discourage waiting. By similar calculation, a 16 percent tax credit in 1990 will discourage waiting beyond that year. Waiting beyond 1989 and 1988 could be discouraged by tax credits of 31 and 40 percent, respectively, in those years. Waiting beyond 1987 could only be discouraged by a whopping 62 percent tax credit, far above the current level (see *Table 4.5*).

The policy implications of this arithmetic are clear. A tax credit at a constant tax level can discourage the early adoption of solar-energy systems. It is debatable whether tax credits should encourage investments in photovoltaics prior to 1990, when they become profitable from a public-policy point of view. The phased reduction of the solar-investment tax credits in the late 1980s, however, would reduce the advantage of waiting for system prices to drop further. Clarifying the price signals this way would discourage both consumers and manufacturers from holding back.

Table 4.6 Population in American Insolation Belts, 1970 and 1980

Belt	kwh/m²	1970 thous.	%	1980 thous.	%	growth rate (%)
A	2.33–3.49	7,087	3.5	7,290	3.3	2.9
B1	3.49–4.07	92,541	45.6	95,313	42.7	3.0
B2	4.07–4.66	44,531	21.9	49,138	22.0	10.3
C1	4.66–5.24	48,635	24.0	58,000	26.0	19.3
C2	5.24–5.82	8,093	4.0	10,550	4.7	30.4
D	5.82–6.98	2,110	1.0	2,834	1.2	34.3
total		203,212		223,125		9.9

Demography and Solar Markets

Our analysis indicates the importance of both insolation and energy costs on the profitability of solar-energy systems. In the United States people and economic activity are shifting toward areas of greater insolation and lower population density (*Figure 4.5*). Although only a small proportion of the population inhabits the areas of greatest insolation, their growth rates have exceeded areas of least insolation by a wide margin (*Table 4.6*). For example, population in the area from El Paso to the Mojave Desert grew 34 percent from 1970 to 1980, compared to a growth of 3 percent in the rainy areas of northern New England and the Pacific Northwest. The continuation of this movement to the Sun Belt suggests that the market for solar-energy systems that utilize insolation directly will grow fastest in the areas where they yield the greatest output. Since many of these areas are heavily dependent upon expensive oil and gas, demographic trends may reinforce the trends in technology and fuel prices that favor the growth of solar markets.

Conclusions

The twin forces of decreasing hardware costs and escalating fuel costs are likely to create significant markets in two applications in the near future: remote irrigation by the mid- to late 1980s and utility-interactive residences by the early 1990s. The higher the fuel-escalation rate and the lower the discount rate, the earlier the viability. The population shift to sunnier regions of the United States will further stimulate the growth of these markets.

The results for irrigation can be generalized to semiarid developing countries in the tropics. Here the benefits of residential systems were taken from existing utility rates. These rates are only a crude reflection of the marginal costs of the conventional electricity displaced. In the next two chapters, the costs that utilities can avoid as a result of renewable energy systems are computed. When benefits are measured by avoided costs, photovoltaics appear to be viable in Sun Belt sites like Fort Worth much earlier than indicated here.

Notes

1. The latter assumption is warranted in regions that depend upon surface water or that have a high water table. The assumption is unwarranted in the semiarid zone above the Ogallala aquifer, stretching from western Nebraska to Texas. See Green, 1973.

2. Data on land values and fuel costs were obtained from interviews at the Mead, Nebraska, Agricultural Experiment Station. For more information, see Matlin and Katzman, 1977, Table III.

3. Arithmetically, $.04t = \ln(\$7 + \$60 - \$27) - \ln(\$7)$; therefore $t = 44$.

4. If a DC pump were used, inverter costs could be saved. DC pumps, however, are more expensive than commonly available AC pumps. Whether powered by a solar or fossil-fuel generator, the pump has a mechanical efficiency of 60 percent.

5. To deliver 1 kwh to the consumer's meter, the utility actually must generate 1.1 kwh. When the consumer uses 1 kwh, the consumer in reality pays for 1.1 kwh. When 1 kwh is sold back to a central power station, the utility receives only 0.9 kwh. The value of 1 kwh received versus 1 kwh delivered by the utility is only 0.9 to 1.1, or approximately 90 percent.

6. The current law provides a $10,000 limit on the outlay eligible for the tax credit.

5
Farming the Wind

Zephyrus, the powerful god who propelled the ships of the ancient Argonauts, drained the Dutch polders, and watered the hardscrabble ranches of west Texas, has been embraced by big business. The commercial exploitation of wind energy appears to be promising. A combination of high average wind speed, high electricity costs, and favorable tax laws has made "wind farming" attractive in a few favored locations today. Zephyrus' resurrection has been heralded as an attractive tax shelter (BW, 26 April 1982).

While wind energy has been converted to mechanical energy for pumping water and grinding grain, the technology for converting it into electricity is the most likely vehicle for increasing its contribution to the world energy supply. Wind-energy conversion systems (WECS) lend themselves to adoption on a stand-alone basis, but there are major advantages of utility interaction. Because demands for electricity and the supply of wind energy are rarely in phase, the utility can serve as a source of backup as well as a sink for surplus electricity.

Several utilities in the United States are beginning to exploit this wind potential. Some have constructed large-scale windmills, while others have contracted to purchase electricity from privately owned "wind farms" (ELP, May 1981, April 1982; SE, August 1981, pp. 26–30; NERC, 1983). For example, Hawaiian Electric Company is contracting to purchase electricity from a wind farm whose capacity is about 7 percent of its own conventional capacity.

The economic value of wind energy to utilities is indicated by the savings in conventional fuel and capacity. These savings influence the likelihood of a utility's installing WECS capacity of its own or the terms it offers to potential wind farmers. These terms will in turn influence the profitability of nonutility ownership of wind farms or WECS dedicated to specific load centers. For the latter, these terms also affect the profitability of storage vis-à-vis selling surplus electricity to the utility.

The Supply of Wind Energy

The physics of wind energy is well known (Golding, 1976; Inglis, 1978; Eldridge, 1980). Broad global wind patterns result from the uneven heating of the atmosphere and the rotation of the earth. At the regional level, the speed and direction of the wind are influenced by land forms such as mountains or valleys and the juxtaposition of water and land. At the local level, small variations in topography or surface friction from vegetation or buildings can influence wind speed. Because of friction with surface features, wind speed increases with altitude, generally following a "one-seventh power" law; that is, as the height of the wind machine hub doubles, wind speed increases by one-seventh.

Slight variations of wind speed with location and altitude have great economic significance. Because wind energy varies with the cube of wind speed, a doubling of wind speed results in an eightfold increase in energy. Twofold variations of wind speed with topography, surface vegetation, or altitude are common within a small region. This cubic-power law implies that there are great payoffs to careful prospecting among alternative sites for wind energy conversion systems.

Not all of the 600 quads of wind energy that brush the surface and the 60 quads that skirt the coastline of the United States are available for human use because of siting limitations, inefficiencies in wind energy conversion, and economic constraints. At the local level, siting limitations are hardly trivial. The minor but adverse environmental impacts of wind machines—interference with television reception,[1] the incessant drone, and the risk of blade ejection—discourage their erection in densely settled areas. Land for wind energy conversion is best obtained in rural areas, where the interference with ongoing activity is minimal. In rural areas like picturesque mountain passes or virgin beaches, aesthetic and recreational values may further restrict wind-farm development. Siting wind machines in rural areas, away from urban load centers, can add the cost of transmission lines to wind-energy systems. These costs would be relatively minor if a wind machine was located near existing transmission lines or near large rural industrial sites. Because only about 3 percent of the nation's land area is urbanized, siting limitations should not be a severe problem on a regional level. Offshore wind machines should face even fewer siting limitations.

The scientific principles underlying wind-energy conversion are also well known. A perfectly efficient wind machine can capture nearly 60 percent of the energy available in the wind within the area swept by its blades. In practice, wind machines fall far below this theoretical limit. The picturesque Dutch windmill, with four sail-like blades, captures about 17 percent of the available wind energy. The multiblade American farm windmill captures about 30 percent. Some modern propeller-type wind machines can capture more than 40 percent of the available wind energy under certain operating conditions, but efficiencies of 35 percent are more common (Eldridge, 1980, pp. 130–37). The conversion of energy from the wind turbine into high-quality electricity is about 90 percent efficient. The combined efficiency of the wind machine rotor and the generator is about 30 percent (35 percent × 90 percent). There appears to be relatively little room for efficiency improvement. Efficiency limitations alone bring the available wind energy down from 660 quads to about 200 quads, a large sum compared to total American energy consumption of about 70 quads.

Wind-energy conversion systems do not follow the cubic-power law precisely. At speeds below the "cut-in speed," wind machines generate no energy. To avoid the risk of blades flying off through centrifugal force, the speed of rotation peaks at a certain "rated speed." Most wind machines shut down entirely in very high winds, above the "cut-out speed." The performance curve for a commercially available machine is plotted in Figure 5.1. This wind machine begins generating electricity at 7 MPH. At speeds between 7 and 30 MPH, power increases at a rate a bit less than predicted by the cubic-power law.[2] In the range of 30 to 60 MPH, the machine can generate 50 kw, its rated capacity. At speeds faster than 60 MPH, the cut-out speed, the machine shuts down.

Since wind machines do not convert all the energy in very high or very low winds, the capturable share of wind energy is reduced even further. Because of the cubic-power law, energy losses at low speeds are relatively minor, but losses at high speeds may be substantial. How much is lost depends upon the distribution of wind speed at a particular site.[3] In sites where wind speed averages 24 MPH, more than half of the energy would be lost by our sample machine (Eldridge, 1980, p. 115). In sites like Kahuku, Hawaii, where wind speed averages 17 MPH, about 3 percent of the energy is lost.

In wind-machine design, there are potential trade-offs between mechanical efficiency, wind-speed limits, and materials costs. An efficient

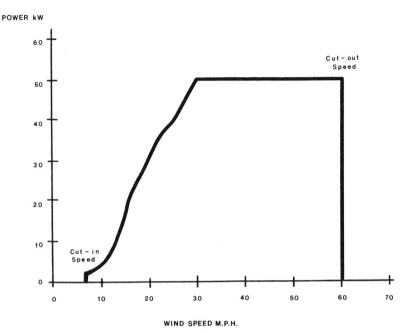

Figure 5.1 Performance curve: 50-kw windmill system. Source: Energy Sciences, Inc., for ESI-54.

wind machine may deliver energy at a higher cost than a less efficient machine. At a higher cost, a wind machine with a lower cut-in and higher cut-out speed could be designed for converting a higher proportion of the wind energy. Several competing engineering issues are yet unresolved, such as the merits of small versus large machines or vertical versus horizontal axes.

Because wind machines differ in their rated speed, it is difficult to compare the costs of different systems on a capacity basis. Currently, small wind-energy conversion systems (fewer than 25 kw) can be installed for about $5,000/kw. Medium-size wind machines (25 to 250 kw) can be installed for about $2,000/kw.[4] Large-scale systems (more than 250 kw) currently cost over $4,000/kw (ELP, May 1982, p. 4). These cost estimates may be substantially higher in sites where soil conditions require special footings. Mass production may cut the costs of large-scale wind machines in half, and improvements in materials may cut costs even further (McGraw, 1981). Engineering studies, which take into account these potentials for improvement, indicate that the ultimate cost per kilowatt is unlikely to vary much within a fairly large range of sizes (Inglis, 1978, appendix 2; Eldridge, 1980, pp. 153–60).

The Economics of the Wind Farm

Less important than the efficiency or cost per unit of wind-machine capacity are the costs and benefits of each unit of energy delivered. The economic viability of a wind-energy conversion system depends upon its capital and maintenance costs and the value of the electricity displaced. Let's consider the economics from the public-policy and investor points of view.

For the production of renewable energy, the 1981 Economic Recovery Act encourages the establishment of third-party partnerships that are neither regulated utilities nor customers. The sole purpose of these wind-farm partnerships is to sell electricity to utilities. A partnership can enjoy a 25 percent tax credit for energy-saving investments and a five-year schedule of depreciation. The credit on the $100,000 cost of a 50-kw wind-energy conversion system reduces the investor's tax liability by $25,000. If the investor is in the 50 percent tax bracket, the present value of the depreciation allowance is about $40,000. Together, the depreciation allowance and tax credits reduce the cost of the project by 65 percent. Offsetting these advantages is the income tax on profits from electricity sold to the utility, in this example a 50 percent tax. Under the tax code prevalent in the 1980s, a utility enjoys only a 10 percent tax credit and must depreciate alternative-energy investments over ten years.

The wind farmer purchases "wind rights," installs a wind system, and contracts to sell electricity to a utility at a given price. Because of the environmental impact, the wind farm is invariably sited in a rural area. Once the system is installed, the wind farmer incurs annual maintenance and operation expenses. The investors bear the risk that electric generation is significantly less than expected and that falling towers or errant blades may damage persons, crops, livestock, or property.

The evaluation of a wind-farm investment must be site-specific because of regional variations in wind-energy potential, in electric-load cycles, and in utility-capacity mix, which engender regional variations in fuel and capacity savings. Five representative sites in the United States are examined here in increasing order of average wind speed (measured at 30 feet): Miami (7.7 MPH), El Paso (8 MPH), Fort Worth (10 MPH), Boston (12 MPH), and Kahuku, Hawaii (17 MPH). Kahuku experiences one of the most favorable wind regimes in the world, while El Paso and Miami experience some of the lowest wind speeds. This range of wind speeds can be observed on most continents.

At each site, the performance of seven commercially available WECS was simulated by Program WINDMILL PERFORMANCE. The ma-

Table 5.1 Annual Output, Capacity Factor, and Levelized Cost of Windmill-Generated Electricity (cents/kwh)

	Kahuku	Boston	Fort Worth	El Paso	Miami
Output Mwh	182	95	62	42	30
Capacity factor	42%	22%	14%	10%	7%
Social cost					
5% discount	5.6	10.5	16.2	24.0	33.4
10% discount	7.7	14.4	22.2	32.9	45.8
Wind-farmer cost					
5% discount	3.8	7.5	11.6	17.3	24.1
10% discount	6.0	11.2	17.2	25.7	35.8

chine with the median yield at wind speeds in the 13 to 15 MPH range is selected for further analysis (*Figure 5.1*). The WECS is assumed to be unavailable 5 percent of the time for forced and scheduled outages. Annual operation and maintenance costs are assumed to be 2 percent of installed cost.

For the Hawaiian site, hourly wind speeds are taken from an average year, the best year in five, and the worst year in five.[5] For the four mainland sites, hourly wind speeds are taken from SOLMET files for "typical meteorological years." These monitored stations are typical and not necessarily optimal from the viewpoint of siting wind machines.

Because gusts may double or triple wind speed in a matter of seconds, the use of hourly data at first seems suspect. Hourly data may indeed provide a poor measure of the performance of a single WECS; however, wind farms are likely to comprise dozens of WECS spread over acres. On the wind farm, the effects of local gusts would be unnoticeable.

The simulations indicate that annual output of the 50-kw WECS ranges from 30 MWh in Miami to about 180 MWh in Kahuku. The variation in output between the best and worst years is less than 10 percent in the Hawaiian site, so year-to-year variations are ignored.

The output of an electric generating system is often indicated by its "capacity factor." This is the ratio between actual output and potential output if the system operated at full capacity throughout the year. This factor ranged from 7 percent in Miami to 40 percent in Kahuku.

The levelized cost of wind energy depends upon the discount rate and upon the perspective of the investor. WECS costing $2,000/kw can yield electricity at a social cost of 6 to 8 cents/kwh in windy Kahuku and at a cost of 33 to 46 cents/kwh in Miami. These levelized costs are about 30 percent lower from the perspective of a wind farmer, who takes into account the tax advantages (*Table 5.1*).

For a private investor, a wind farm is profitable if the price offered by the utility is greater than the investor's cost. According to PURPA guidelines, the price that a state regulatory commission requires a utility to pay is supposed to reflect the utility's avoided fuel and capacity cost. In their contracts with wind farms, utilities tend to assume that capacity savings will be minimal, but that fairly expensive fuels, such as oil or gas, will be saved, in addition to minor maintenance expenses. In Hawaii, a utility is offering 5 cents/kwh for fuel savings and will consider offering a premium for capacity savings after about five years of operating experience. In California, a utility is offering 11 cents/kwh for the first three years of wind-farm operation.

Avoided costs may be higher or lower than the rates actually allowed or mandated by utility commissions. Fuel savings vary with the source of energy displaced, which depends upon when the energy is delivered over the load cycle. During peak periods, utilities tend to burn expensive gas or oil in thermodynamically inefficient plants. During the slack base periods, coal or nuclear fuels are commonly used. The value of these fuel savings can vary by a factor of 10 during the course of the day. Because service interruptions are most likely when the load is at the peak, wind energy delivered during the utility's peak can contribute to system reliability and reduce requirements of conventional capacity. The closer the fit between the peak load and the peak wind-energy cycles, the greater the contribution of WECS to both fuel and capacity savings. The match between patterns of daily load and daily wind energy differ substantially from one utility to the next and from season to season.

Because of the lack of field experience, utilities and potential wind farmers must draw upon other sources of information about avoided fuel and capacity costs. A promising method of evaluating these effects is computer simulation. Simulation involves modeling an electric utility as a set of equations that describe the important engineering parameters of the utility, as well as behavioral rules of system operators and investment planners. These models predict how much it will cost utilities to meet their loads from a given capacity mix and how reliable this capacity will be. Several large-scale computer simulations have been developed to assess the impact of renewable energy systems on utilities (Giese, 1979; Finger, 1980).

How Valuable Are the Fuel Savings?

Fuel savings can be evaluated by a utility production-costing model, which simulates the dispatching of conventional generating capacity. This study uses Program GENCOST, a mirocomputer variation of these models that has been successfully validated (Katzman and Katzman, 1982a, 1982b, 1983). In the presence of a renewable energy source, production costing requires the following key inputs: (1) estimates of utility loads, (2) estimates of performance of the renewable energy system, (3) estimates of conventional generating capacity, (4) assumptions about the performance of conventional generators, (5) rules for dispatching available generators, and (6) economic parameters. The most interesting output is the total fuel cost at different levels of WECS capacity. The change in total fuel costs that result from increasing WECS capacity provides a measure of the value of depletable energy that windmills save.

The impact of wind-energy conversion systems is estimated for five utilities: Hawaiian Electric Company, which serves the island of Oahu; El Paso Electric, a metropolitan utility; Texas Utilities, whose transmission lines sweep across the northern part of the state, centering on Dallas/Fort Worth; Florida Power and Light, which serves most of the eastern coast and southern half of the state; and Boston Edison, another metropolitan utility. These utilities are typical of those having a heavy dependence upon generators that burn expensive oil or gas.

Load Cycles

Electric loads follow a daily rhythm that is harmonized with human activity and the weather. In all regions, electric loads are relatively low in the spring and fall. In the winter, loads peak in the morning as people rise to prepare for work or school. Loads peak again in the evening as the members of the household return for dining and recreation. Electric loads of heavy industry are generally fairly uniform around the clock. On the average, patterns of wind energy show less cyclical variation. Average load cycles in Boston and Fort Worth cover the typical range across the nation (EPRI, 1977). In these sites, wind energy peaks in the afternoon in all seasons (*Figures 5.2* and *5.3*).

Production-costing models generally assume that the load cycles are given, although they may change over the life of the wind-machine investment. Consumer conservation is expected to reduce demand proportionately in all hours (OTA, 1980, Table 62). Load-management practices like time-of-day pricing can be expected to reduce the peaks in demand. Changing lifestyles, such as the continued entry of women into the labor force, may reduce daytime electric loads and increase the morning and evening peaks. Because the net effects of these changes are hard to forecast, load data for the 1979–81 period are taken as representative of the future. This assumption of load stability is a matter of convenience rather than necessity. It permits the extrapolation of the results of a single year to the twenty-year assumed lifetime of the WECS, instead of performing twenty annual simulations.

Most utilities report average hourly load data in the standard format

92 *Solar and Wind Energy*

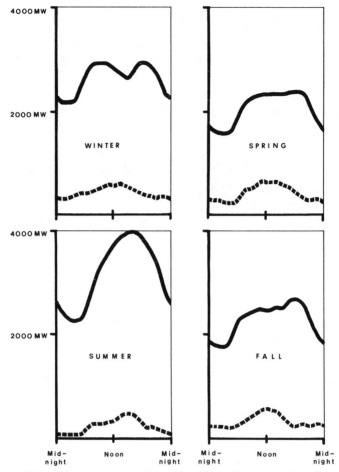

Figure 5.2   Average daily profiles, Fort Worth: Loads and windmill output. (Loads indicated by solid lines; windmill output, by dotted line.)

of the Edison Electric Institute. Average-load data hide momentary power surges and valleys, but these are relatively minor.

The Performance of WECS

For all practical purposes, the analysis is independent of whether the WECS are dispersed in many small wind farms or in a few large-scale power stations. The output for the entire wind farm is simply based on scaling up the wind speed-performance curve of the solitary 50-kw WECS. The minor losses that result from the clustering of wind turbines

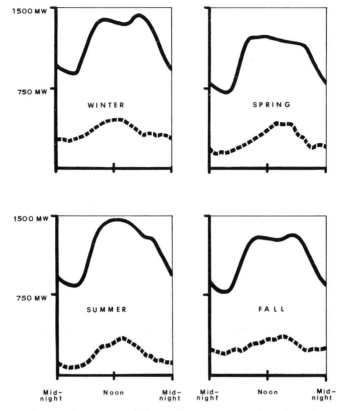

Figure 5.3   Average daily profiles, Boston: Loads and windmill output. (Loads indicated by solid lines; windmill output, by dotted line.)

is ignored. The four levels of WECS capacity examined represent 3, 10, 20, and 30 percent of peak load on the utility. The output of the wind farm is subtracted from the original no-WECS load to create a new load profile.

Conventional Generating Capacity

Forecasting the generating capacity of utilities over the assumed twenty-year lifetime of a wind farm is relatively easy. Because of the longevity of the existing capacity and the prolonged lead time for new construction, capacity can be projected as far as seven years ahead with some confidence. Forecasts of the 1990 generating mix are taken from company annual reports and compilations of the North American Electric Reliability Council (1982a). The council provides ten-year projections of generating capacity of the power pools around the country. Since utilities are generally interconnected, the projected mix of the power pools may be as relevant as the mix projected by the company itself. The two projections of capacity mix are quite close, but company forecasts are preferred because they offer more detail about specific generating units. Because of the slowdown in new construction, this mix of plants is assumed to be stable over the relevant time horizon. Again, this assumption of stability is a matter of convenience rather than a necessity.

The Performance of Conventional Generators

For purposes of simplicity, the following assumptions are made about the operating characteristics of electric utilities:

First, the heat rate of all base-load and cycling units is taken as 10,000 Btu/kwh, which is close to the national average. In fact, new efficient boilers can have heat rates as low as 9,000 Btu/kwh, while old ones have rates as high as 12,000 Btu/kwh. Taking the average distorts the results only slightly. Small combustion turbines, which are used for peaking, have assumed heat rates of 13,000 to 14,000 Btu/kwh.

Second, routine maintenance of conventional generators is scheduled to minimize the risks of interruption in service. Maintenance may take several weeks, so the best periods for scheduled maintenance are those with low weekly peak loads. In practice, maintenance is usually scheduled for spring and fall, when cooling and heating demands are low. In the simulations, annual maintenance is scheduled for eight weeks for nuclear, six weeks for coal, and two weeks for oil and gas units.

Third, the forced outage rate of a unit depends upon its fuel, size, and age. The probability of forced outage averages about .15 for nuclear, .13 for coal, and .05 for oil and gas units. Rates are higher for larger units. Over time, rates may decline during the initial break-in period, then rise gradually as the unit approaches retirement. For simplicity, the average forced outage rates noted above are used. The simulation "derates" generating capacity by the forced outage rate. In other words, a 100-MW coal plant with a forced outage rate of .13 is treated as a fully available 87-MW plant (General Electric, 1978, vol. 3, appendix 3; Finger, 1980).[6]

Dispatching Rules

The available capacity is dispatched in order of increasing fuel and operating costs per megawatt-hour. Because of their relatively low heat rates and fuel costs on an MBtu basis, nuclear or coal units are dispatched first. Oil or gas boilers are dispatched next, and gas or oil turbines, which have fairly high heat rates, are dispatched last.

Economic Parameters

The major economic parameters that affect the value of fuel saved by wind-energy conversion systems are the costs of the various fuels, fuel-escalation rates, and the discount rate.

The replacement cost of low-sulfur residual fuel oil in 1980 was about $6/MBtu. Natural gas is imputed a marginal cost of $6/MBtu in the producing states of the Southwest, where it is used as a utility fuel. The replacement cost of coal is taken as $1/MBtu in the Southwest and $2/MBtu in the Northeast. The marginal cost of nuclear fuel is taken as $1/MBtu nationally. At average heat rates of 10,000 Btu/kwh, each $1/MBtu translates into 1 cent/kwh. External costs of these fuels are considered later.

Most maintenance costs vary with the number of hours that capacity is utilized during the year. Coal and nuclear plants incur additional costs of maintaining large fuel inventories. Schurr et al. (1979, Table 9-6) estimate operation, maintenance, and fuel inventory costs at 22 mills/kwh for coal and nuclear plants. Finger (1979, pp. 165–66) estimates operation and maintenance costs at 2 mills/kwh for oil and gas plants. Thus 2 mills/kwh appears to be a reasonable estimate for all plant types. These relatively minor costs are not escalated on the assumption that rising wages for maintenance will be compensated by productivity increases.

Fuel Savings

Simulation results are presented for the five sites at the four levels of windmill capacity. For example, in Miami, peak load is about 8,800 MW and the corresponding levels of windmill capacity about 265, 880, 1760, and 2540 MW. To facilitate comparisons among utilities of different sizes, the savings are standardized on the basis of a kilowatt of windmill capacity.

Differences in wind speed produce substantial differences in the output of the wind farms (*Table 5.2*). Each kilowatt of wind capacity generates about 3,800 kwh a year in Hawaii, nearly twice that of Boston, three times that of Fort Worth, four times that of El Paso, and five times that of Miami.

In Hawaii, oil is basically the only generator fuel used, so windmill penetration has no substantial effect on the mix of fuels. In the remaining sites, a disproportion of the savings falls upon the expensive fuels.

Table 5.2 Kilowatt-Hours of Electric Generation Annually Displaced per Kilowatt of Windmill Capacity

	Windmill Penetration as % of Peak Load			
	3	10	20	30
Kahuku				
All oil	3835	3835	3835	3835
Boston				
Nuclear	252	306	380	630
Oil	1788	1736	1663	1597
Gas	3	1	1	0
total	2043	2043	2043	2043
Fort Worth				
Nuclear	0	0	0	0
Coal	604	650	733	809
Gas	752	707	623	547
total	1356	1356	1356	1356
El Paso				
Nuclear	149	169	200	246
Coal	88	90	102	112
Gas	727	705	660	605
total	963	963	963	963
Miami				
Nuclear	0	0	0	1
Coal	37	46	56	69
Oil	674	665	655	641
total	711	711	711	711

For example, about 55 percent of the projected generating capacity of Boston Edison consists of oil-fired boilers, but more than 70 percent of the fuel savings is attributed to these boilers. The greater the windmill penetration, the relatively smaller the share of expensive fuels saved. In Boston, the proportion of the reduction borne by oil-fired generation falls from over 80 percent to less than 70 percent as windmill penetration increases from 3 percent to 30 percent. In Miami, oil-fired generators bear most of the reduction in loads, but the oil share decreases and the coal share increases with greater windmill penetration.

Economic Implications

The economic value of fuel savings can be expressed in two equivalent ways. The discounted present value of the fuel savings can be compared to the capital cost of the WECS, assumed to be $2,000/kw. Or the levelized value of fuel saved per kilowatt-hour of wind energy can be compared to the levelized cost of the WECS.

Table 5.3 Levelized Cost and Savings of Windmill-Generated Electricity (cents/kwh)

	Kahuku	Boston	Fort Worth	El Paso	Miami
Social cost					
5% discount	5.6	10.5	16.2	24.0	33.4
10% discount	7.7	14.4	22.2	32.9	45.8
Wind-farmer cost					
5% discount	3.8	7.5	11.6	17.3	24.1
10% discount	6.0	11.2	17.2	25.7	35.8
1980 value of fuel					
3% penetration	6.2	5.4	3.8	4.8	5.8
30% penetration	6.1	4.9	3.0	4.1	5.6
Levelized value of fuel (2% fuel/10% discount)					
3% penetration	7.1	6.2	4.4	5.5	6.7
30% penetration	7.0	5.6	3.5	4.7	6.4
Levelized value of fuel (3% fuel/8% discount)					
3% penetration	8.1	6.9	4.8	6.1	7.4
30% penetration	8.1	6.2	3.8	5.2	7.1
Levelized value of fuel (4% fuel/5% discount)					
3% penetration	11.7	10.2	7.1	9.0	10.9
30% penetration	11.5	9.2	5.6	7.7	10.5

In Hawaii, WECS save oil, which was worth about 6.2 cents/kwh in 1980. In Miami, the value of the fuel mix is 5.8 cents/kwh at low penetrations and 5.6 cents/kwh at high penetrations. In Boston, the value of the fuel mix falls from 5.4 cents/kwh to 4.9 cents/kwh with increased wind penetration. At the low end, the value of fuel saved in Fort Worth is only 3.8 cents/kwh at low penetrations.

Because the price of fuels is expected to escalate above inflation, the levelized value of the fuel savings is higher than the current value. If the expected escalation and discount rates are 3 and 8 percent, respectively, then the levelized value of the fuel savings is nearly 30 percent higher than the 1980 value (*Table 5.3*).

Except for the low fuel-escalation/high discount-rate scenario, the levelized value of fuel savings exceeds the levelized social cost of wind energy in Kahuku. In all other sites, social costs exceed the value of fuel savings. In Hawaii, the value of fuel saved clearly exceeds the wind farmer's levelized cost of wind energy at the discount and fuel-escalation rates considered. In Fort Worth, El Paso, and Miami, the wind farmer's levelized cost of wind systems exceeds the value of the fuel saved at all rates considered. In Boston, the balance between the wind farmer's levelized cost and fuel savings is sensitive to discount and fuel-escalation rates.

The discounted present value of the fuels and operating costs saved by windmill capacity are computed at three discount (5, 8, and 10 percent) and three fuel-escalation rates (2, 3, and 4 percent). Calculations are undertaken for systems installed in 1980, 1985, and 1990. All calculations are based upon the public-policy calculus and thus exclude tax considerations. The most illustrative scenarios reflect the best case for wind power (high fuel escalation/low discount), the worst case (low fuel escalation/high discount), and the most likely case (moderate fuel escalation and discount). The range of results are exemplified in Hawaii, Boston, and Miami (*Table 5.4*):

1. As expected on the basis of both windmill output and fuel costs, the value of fuels saved is the highest in Hawaii and the lowest in Miami. In Hawaii, fuel savings under all scenarios and penetrations exceed by far the $2,000/kw cost of the wind-energy conversion system. In Miami, fuel savings under all scenarios are far below these system costs. In Boston, fuel savings are close to $2,000/kw costs under the best-case scenario.
2. The value of fuel savings per kilowatt of WECS diminishes with penetration. This follows from the fact that increased penetration saves increasingly cheaper fuels used in base- and intermediate-load generating plants.
3. Because fuel costs are expected to rise faster than inflation, WECS installed in later years will enjoy a higher present value of fuel savings. If the costs of WECS also decline over time as the industry achieves economies of scale and experiences technological improvements, the economic prospects for installing utility-interactive WECS will become even more favorable.
4. Savings in maintenance and operation of conventional plants are inconsequential at all sites and at all penetrations.

If the external costs of fuel are taken into account, the picture does not change dramatically. An external cost surcharge of, say, 25 percent raises the discounted social benefits of WECS by a corresponding amount. For sites with a wind regime like Fort Worth's, even a 100 percent surcharge cannot push savings above costs. For a wind regime like Boston's, a surcharge as small as 25 percent can push the WECS within striking distance of viability by 1985. For example, at 3 percent penetration under the worst-case scenario, such a surcharge raises net benefits from less than $1,200/kw to about $1,500/kw.

These conclusions hardly change when tax credits and depreciation allowances are considered. These tax advantages reduce the present

Table 5.4 Present Worth of Fuel Savings ($/kw) per Kilowatt of Windmill Capacity: Alternative Scenarios

	Wind Penetration as % of Peak Load			
	3	10	20	30
Kahuku	Worst case: 2% fuel/10% discount			
1980 installation	2290	2278	2257	2244
1985 installation	2581	2568	2545	2530
1990 installation	2796	2782	2757	2741
M&O	63	63	63	63
Capacity	498	348	269	203
	Most likely: 3% fuel/8% discount			
1980 installation	2946	2931	2904	2887
1985 installation	3527	3509	3477	3457
1990 installation	3976	3956	3920	3897
M&O	73	73	73	73
Capacity	498	348	269	203
	Best case: 4% fuel/5% discount			
1980 installation	4309	4287	4249	4224
1985 installation	5478	5450	5401	5369
1990 installation	6429	6396	6338	6301
M&O	95	95	95	95
Capacity	498	348	269	203
Boston	Worst case: 2% fuel/10% discount			
1980 installation	1055	1028	993	961
1985 installation	1189	1160	1120	1084
1990 installation	1288	1256	1213	1174
M&O	32	31	30	30
Capacity	419	323	267	230
	Most likely: 3% fuel/8% discount			
1980 installation	1518	1452	1366	1289
1985 installation	1764	1687	1588	1498
1990 installation	2049	1961	1844	1739
M&O	37	36	36	35
Capacity	419	323	267	230
	Best case: 4% fuel/5% discount			
1980 installation	2220	2125	1999	1886
1985 installation	2767	2647	2490	2349
1990 installation	3313	3170	2982	2813
M&O	47	47	46	45
Capacity	419	323	267	230
Miami	Worst case: 2% fuel/10% discount			
1980 installation	394	391	387	382
1985 installation	445	441	436	431
1990 installation	482	478	473	467
M&O	12	12	12	12
Capacity	60	36	48	36
	Most likely: 3% fuel/8% discount			
1980 installation	507	503	498	491
1985 installation	608	602	596	588
1990 installation	685	679	672	663
M&O	14	14	14	14
Capacity	60	36	48	36
	Best case: 4% fuel/5% discount			
1980 installation	742	736	729	719
1985 installation	944	936	926	914
1990 installation	1107	1098	1087	1072
M&O	18	18	18	18
Capacity	60	36	48	36

value of system costs by about 65 percent, to about $700. But the 50 percent marginal personal income-tax rate (paid by the hypothetical wind-farming partners) reduces the value of $2,000 in fuel savings to $1,000. Because taxes reduce costs more than benefits, the wind farm looks relatively more attractive on the balance sheet of the wind farmers than on the public balance sheet.

Generalizing the Results

In only one of the sites examined (Kahuku) was a wind-energy conversion system clearly economic under all plausible fuel-escalation and discount-rate scenarios. In this site, two favorable factors were joined: high wind speed and dependence of the utility on high-cost oil.

To generalize to other sites, we calculate the levelized cost of wind-generated electricity and the fuel savings for four ideal-type utilities: (1) an all oil-burning utility (like Hawaiian Electric); (2) a utility in which 75 percent of the savings are in oil or gas and the rest in coal or nuclear (like Boston Edison); (3) a utility in which about half of the savings are in oil or gas and the rest in coal or nuclear (like Texas Utilities or El Paso Electric); and (4) an all-coal/nuclear utility, like those in midcontinental North America.[7] The output of the particular wind-energy conversion system has been simulated by the manufacturer for average wind speeds between 12 and 22 MPH.

At these wind speeds, the levelized cost of wind energy ranges from 4.8 cents/kwh to 10.1 cents/kwh. At a modest fuel-escalation rate, the value of fuel savings is 8.4 cents/kwh for the all-oil utility, 6.9 cents for the 75 percent oil/gas utility, 4.8 cents for the 50 percent oil/gas utility, and less than 2.0 cents/kwh for the all-coal/nuclear utility (*Table 5.5*).

A comparison of levelized costs and benefits of wind systems indicates that they are competitive with electricity generated totally with oil at sites with average wind speeds as low as 14 MPH. Wind systems are competitive with 75 percent oil-generated electricity at average wind speeds as low as 16 MPH. When only 50 percent of the fuel savings is in oil or gas, windmills are competitive in sites with average wind speed above 22 MPH. Windmills are far from competitive with coal- or nuclear-generated electricity, even at sites with that higher wind speed.

Another way of analyzing these results is to compute the cost that wind systems would have to be in order to compete at sites with lower wind speeds. For example, to be competitive with 50 percent oil/gas-generated electricity at 12 MPH sites, the installed cost of wind systems would have to decline by more than 50 percent, to less than $1,000/kw. The decline would have to be about 80 percent to render wind systems competitive with all-coal/nuclear-generated electricity. In sites with higher wind speeds, the required declines would have to be correspondingly less.

Table 5.5 Levelized Social Cost and Savings from Wind-Generated Electricity: 3% Fuel Escalation/8% Discount Rate (cents/kwh)

Wind speed mph	Levelized cost	Utility Type			
		all oil	75% oil/gas 25% coal/nuclear	50% oil/gas 50% coal/nuclear	all coal/ nuclear
12	10.1	8.1	6.9	4.8	2.0
14	7.7	8.1	6.9	4.8	2.0
16	6.3	8.1	6.9	4.8	2.0
18	5.5	8.1	6.9	4.8	2.0
20	5.0	8.1	6.9	4.8	2.0
22	4.8	8.1	6.9	4.8	2.0

Note: Annual kwh simulated from manufacturer's field performance data and standard Rayleigh distribution.

Source: Energy Sciences, Inc.

Finally, these results provide some indication of allowable transmission costs. Suppose that a city served by a 75 percent gas/25 percent coal utility had a local wind speed of 14 MPH. A wind farm at this location would be unprofitable because levelized costs of 7.7 cents/kwh exceed levelized fuel savings of 6.9 cents/kwh. If this city could be linked to a site with 16 MPH winds at a levelized transmission cost of less than 0.6 cents/kwh, then this remote wind farm would be profitable.

Are There Any Savings in Conventional Capacity?

In planning their capacity requirements, utilities take into account the inevitability of equipment failures. Planned capacity must exceed the expected peak load by a margin of safety. Traditionally, this reserve margin has been set by rule of thumb, such as some percent of peak load. Increasingly, utilities are considering the risks of failing to meet loads at any time, not just during the system peak. The capacity savings from wind farms are estimated first by the traditional approach and then by the newer one.

Impacts on Peak Loads and Load Factors

An indicator of capacity utilization, the "load factor" is the ratio of average load to peak load. As wind-generating capacity increases, the average load on all utilities obviously decreases. The peak load on the Boston and Fort Worth utilities decreases somewhat, but the peak load

Table 5.6 Peak and Average Loads (MW) and Load Factor at Different Levels of Windmill Penetration

% windmill penetration	Peak load	Average load	Load factor
Hawaii			
0	917	710	.774
3	916	697	.761
10	916	669	.731
20	916	629	.687
30	916	589	.643
Boston			
0	2100	1203	.573
3	2083	1188	.570
10	2046	1154	.561
20	2010	1105	.549
30	1997	1056	.529
Fort Worth			
0	11202	6183	.552
3	11151	6131	.550
10	11107	6010	.541
20	11107	5836	.525
30	11107	5663	.510
El Paso			
0	688	416	.604
3	688	413	.600
10	688	408	.593
20	688	400	.582
30	688	393	.571
Miami			
0	8860	5169	.583
3	8860	5148	.581
10	8860	5097	.575
20	8860	5025	.567
30	8860	4954	.559

on the Hawaii, El Paso, and Miami utilities shows virtually no decrease at all. In all five sites, increased wind capacity results in lower load factors. How can this be? The explanation is that in Hawaii, El Paso, and Miami winds are below the cut-in speed during some hours of near-peak load on the utility (*Table 5.6*).

Under the traditional planning approach, wind-energy conversion systems appear to save no capacity in Hawaii, El Paso, and Miami. How large is the potential capacity savings in Boston? Ten percent of peak load is 210 MW. If WECS capacity were this large, the peak load would decrease by only 54 MW. But capacity displacement occurs at a diminishing rate. The second 210 MW of WECS capacity saves only 36 MW, the third 210 MW only 13 MW. In other words, at low levels of WECS penetration, it takes 4 MW of WECS to decrease peak load by 1 MW in Boston. At high levels of WECS penetration, it takes 20 MW of windmills to have the same effect. Similarly, in Fort Worth, it takes 12 MW of WECS capacity to reduce peak load by 1 MW at low levels of penetration. No additional peak-load reduction occurs at higher levels of WECS capacity. If capacity requirements are geared to peak load, then WECS generally save little conventional capacity in Boston or Fort Worth.

Reliability and Capacity Savings

The reliability of an electric utility refers to the likelihood that generation, transmission, and distribution capacity can meet loads placed upon it. Attention here is focused upon the reliability of generating capacity. In a conventional utility system, there are two threats to reliability. On the supply side, components of the system undergo random failures or outages. On the demand side, loads fluctuate in predictable cycles with some random variation as well. Consequently, there is some finite probability that loads may exceed capacity at any time. In utility jargon, this risk is called the "loss-of-load probability," or LOLP.

Reliability can be increased by several mechanisms: by increasing reserve margin or capacity above anticipated peak loads, by more frequent preventive maintenance, by spreading risks through additional interutility connections, or by designing systems with additional redundancy. These mechanisms reduce the likelihood that outages (component failures) will result in interruptions (service failures). None of these mechanisms can be pursued without attention to cost.

The installation of utility-interactive WECS imposes new sources of unreliability upon utilities for several reasons. First, while average hourly wind speed is characterized by daily and seasonal cycles, there are substantial random variations around these averages. Second, components of WECS are subject to risks of outage just like conventional systems. The risk of degradation or interruption of performance can also be reduced by installing additional capacity, preventive maintenance, or design redundancy. When there are large numbers of windmills operating in a wind farm, the independent risks of outage are adequately treated by derating, as was done above.[8] The risks of outage of conventional generating capacity is treated differently in loss-of-load calculations.

Loss-of-Load Probability Calculations

The loss-of-load probability of a utility system is customarily measured by the expected frequency of capacity deficiency, such as one day in ten years. If loss-of-load probability is unacceptably high, capacity has to be expanded. If it is too low, some inefficient capacity may be decomissioned or expansion plans deferred.

Most of the steps in calculating the loss-of-load probability are similar to those taken for production costing. The major difference lies in the treatment of scheduled and forced outages. In the loss-of-load model, scheduled maintenance is treated as a one-for-one increase in system load (Vardi and Avi-Itzhak, 1981).

Occurrence of Forced Outage

In the production-costing model, forced outages were treated by derating the equipment. The LOLP calculation recognizes that at any time a given unit is in either an "up" or "down" state. In a generating system with n units, there are 2^n states possible. Given the forced outage rate of each generating unit, the likelihood of each of the 2^n system states is calculated. Next, the events are ordered in increasing magnitude of outage. Finally, a cumulative capacity outage curve can be drawn.

Risks of Interruption

The risks of interruption depend upon the relationship between instantaneous load and available capacity. Hourly loads reflect systematic as well as random components. Available capacity depends upon total system capacity, capacity under maintenance, and the forced outage curve of the remaining capacity.

The LOLP for each hour is the likelihood that available capacity will be insufficient to meet the load for that hour. Because forced outages usually last for at least a day, the daily peak is more relevant to loss-of-load calculations. The loss-of-load probability for the year is simply the sum of these daily probabilities.

Table 5.7 Kilowatts of WECS Capacity Required to Displace 1 Kilowatt of Conventional Capacity

Penetration	Kahuku	Boston	Fort Worth	El Paso	Miami
3%	1.6	2	2.5	3	12
10%	2.7	3	6	3	20
20%	3	3	9	5	15
30%	4.5	3.5	9	6	20

Reliability Results

The numerical results are produced from Program LOLP (see Appendix). As before, the loss-of-load probabilities are calculated for the no-WECS baseline and for four levels of WECS penetration. The LOLP is also computed for several levels of utility capacity in order to determine the savings in capacity that WECS penetration will permit, without increasing the loss-of-load probability. In these simulations, capacity reductions fall upon all conventional units proportionately, although a particular type of unit, such as nuclear or oil, could be made to bear these reductions.

Conventional capacity is reduced in successive simulation runs. One can solve for the level of WECS capacity that is just sufficient to offset the reduction in conventional capacity. Invariably one kilowatt of WECS capacity saves less than one unit of conventional capacity (*Table 5.7*). Two conclusions can be drawn about capacity savings:

1. With the exception of Miami, increasing WECS penetration results in diminishing-capacity savings. The diminishing returns in capacity savings parallel the diminishing returns in fuel savings.
2. Capacity savings are not invariably greater in the windiest sites. At high penetrations, capacity savings in Boston exceed those in Kahuku.

If WECS replace $600/kw oil plants, the value of the capacity credit ranges from $360 per kilowatt of windmills at low penetrations to $180 per kilowatt at high penetrations in Hawaii. If $275/kw gas boilers are replaced in El Paso, the savings range from $140 to $120. These savings are indicated in the last line of each panel in Table 5.4.

The difficulty of saving oil and gas capacity in practice must be emphasized. American utilities routinely buy and sell shares of their generating capacity, but the utility industry as a whole can reduce capacity of a given generating type only by retirement. Because most utilities have more oil and gas capacity than the optimum, it is unlikely that a utility could find a buyer.

At first glance, a more promising alternative to early retirement of oil and gas capacity would be curtailed construction of coal and nuclear capacity. The LOLP simulation was replicated by assuming reductions in nuclear rather than oil or gas capacity. The economics of this strategy was rather poor. Total generating costs shot up dramatically, and the value of fuel saved by windmill capacity dropped. The same result follows from attempting to save coal-fired capacity. This is because the reduction of coal or nuclear capacity results in greater burning of expensive gas and oil and less burning of coal or nuclear fuel. The increased fuel costs resulting from replacing coal or nuclear capacity with a mix of windmill, oil, and gas capacity more than offset the savings in capital costs. Using WECS to avoid having to build coal or nuclear plants appears to be economically unsound. Because the costs of most oil and gas capacities are already sunk, capacity savings may be more chimerical than real.

The Impact on the Volatility of Hourly Loads

Electric loads fluctuate in fairly predictable daily, semiannual, and annual cycles.[9] Superimposed on these predictable or systematic cycles are deviations that result from random variations in contingencies, like the weather. If the behavior of the wind were erratic, then the integration of wind farms into a utility grid might increase the volatility of loads on conventional generators. The greater the volatility of electric loads, the harder it is for utilities to dispatch their existing capacity and to schedule maintenance. Are wind farms likely to have such adverse effects?

If daily loads are subject to great fluctuations, then utilities must keep considerable generating capacity ready for a "warm start." If these daily fluctuations are unpredictable, then much of this capacity must be kept running all the time as "spinning reserve." The utility would obviously save fuel if loads were less volatile and fewer generators had to be kept warm and spinning.

The semiannual or seasonal cycles reflect the swings between mild and extreme weather. Annual load cycles reflect global cooling and heating. Utilities take advantage of seasonal and annual variations in load to schedule maintenance. Obviously, maintenance is scheduled for seasons of low demand. Unpredictable variations from the smooth seasonal cycles make scheduling risky. An unseasonably high demand could catch the utility with its plants down, a financially embarrassing situation. The consequences could either be a brownout or a costly purchase from

Table 5.8 Fluctuations in Utility Loads under Various Windmill Penetrations

% windmill penetration	Average load MW	Daily amplitude MW	Standard deviation MW	Autocorrelation		
				1-hour	2-hour	3-hour
Hawaii						
0	710	185	71	.87	.62	.31
3	697	185	71	.87	.62	.31
10	669	183	75	.88	.65	.37
20	629	183	87	.89	.70	.48
30	589	183	103	.90	.75	.59
Boston						
0	1203	293	198	.95	.84	.72
3	1188	288	199	.95	.84	.72
10	1153	277	205	.95	.84	.72
20	1105	261	222	.93	.83	.72
30	1056	245	248	.92	.83	.72
El Paso						
0	416	97	54	.92	.76	.58
3	413	96	54	.92	.76	.58
10	408	94	55	.92	.76	.57
20	400	92	59	.91	.75	.58
30	393	89	65	.89	.75	.58

another utility. Compared to daily cycles, the semiannual and annual cycles are relatively small and unaffected by windmill penetration.

The amplitude of the daily cycles is examined for various penetrations of windmill capacity in the three sites (*Table 5.8*). The daily amplitude reflects the average rise and fall of electric loads around the yearly average in the course of the day. Let's consider the baseline case, without any windmill capacity. For example, in Hawaii, daily loads average as much as 185 MW above and below the yearly average of 710 MW. In other words, the average load in the peak hour of the day is about 25 percent above the average for the year. These cycles are symmetrical, so at hours when electric demand is at a minimum, the load falls about 25 percent below the yearly average. By coincidence, the amplitude of the daily load cycles are close to 25 percent of the yearly averages in Boston and El Paso as well.

Increasing windmill capacity reduces the amplitude of the daily load cycle in all three sites. Because the average load decreases at about the same rate as the amplitude, the relative volatility of the daily load cycle remains fairly stable. In Boston, for example, the daily amplitude without any windmills is slightly below 25 percent of the yearly average. As WECS capacity increases to 30 percent of peak load, the amplitude falls to about 23 percent of the yearly average. In El Paso, there is also virtually no change in the amplitude relative to the yearly average. In Hawaii, as WECS capacity increases, amplitude rises from 26 to 31 percent of average load. In sum, increased windmill penetration does not affect substantially the systematic portion of the hourly fluctuations in load.

Random deviations in hourly loads from the systematic load cycles are measured by the standard deviation. Under the baseline situation, the standard deviation of the hourly loads is 71 MW in Hawaii. This means that there is one chance in three that loads will be more than 71 MW above or below the load expected on the basis of the systematic cycles. There is one chance in ten that loads may be more than 142 MW above or below expectation. In all three sites, the magnitude of this unpredictable variation in hourly loads increases absolutely, and of course it increases relative to the yearly average. This increased randomness would appear to make the dispatcher's job more difficult.

These hourly deviations in load cycles are not totally unpredictable, however. Hours with especially high loads tend to be followed by hours that also have especially high loads. The tendency of these hourly deviations to be related can be measured by "autocorrelation coefficients."[10] For example, in the baseline case in Hawaii, the one-hour autocorrelation coefficient indicates that about 75 percent ($.87 \times .87 = .75$) of the deviations in the next hour's load can be predicted from this hour's deviation from the average cycles. These coefficients tend to fall off, meaning that it is easier to predict the load one hour ahead than two hours ahead, and so forth. The correlations are still reasonably high five hours ahead, which is approximately the time required to warm up a boiler from a cold start. In all sites, increased wind capacity leaves the autocorrelations virtually unchanged. This means that a dispatcher's job does not become more difficult.

Conclusions

Wind farms are likely to generate economically significant fuel savings for electric utilities today and even more so in the future. Under existing costs of wind-energy conversion systems, sites with wind speeds of 14 MPH are clearly economical on the basis of fuel savings alone if oil or gas is displaced. In these sites, electricity can be generated from the wind at a levelized cost of about 6 to 8 cents/kwh from a social cost-accounting perspective and about 4 to 6 cents/kwh from the point of view of a commercial wind farmer.

As the proportion of gas or oil in the utility fuel mix decreases, higher average wind speeds are required to assure the economic viability of wind-energy conversion systems. In sites served by utilities with considerable coal or nuclear capacity, the wind systems are too expensive by a factor of about three at sites with 18 MPH average wind speeds. As utilities around the world continue to shift from oil to other sources of energy, the profitability of wind-energy conversion systems at current costs decreases. This is because the cost of the fuel displaced declines for most utilities. This means that manufacturers of wind-energy conversion systems cannot simply count on rising oil or gas prices to create a market for their product. The shift to coal by energy users in response to rising oil prices means that manufacturers have to realize significant cost reductions simply to stay even.

These conclusions are hardly altered by taking capacity savings into consideration. In some sites, 1 kilowatt of conventional capacity may be saved by 2 to 5 kilowatts of WECS. In other sites, as many as 20 kw of WECS capacity are required to displace one kilowatt of conventional capacity. While it is hard to generalize about capacity savings, it is clear that compared to fuel savings, they are relatively small for oil- and gas-based utilities.

The simulation methods presented here are more general than the conclusions. The conclusions refer to "typical sites" in various regions and not necessarily the best sites for WECS. Careful wind prospecting will surely uncover profitable sites in all of the regions considered. The most favorable site for a wind system may be in a remote mountain pass or offshore. Significant investments in transmission facilities may have to be incurred to exploit such sites. The rational exploitation of the world's enormous wind energy entails both careful "wind prospecting" and economic analysis.

Whether wind energy can compete beyond the transition from an oil- to a coal-based economy depends on how successful manufacturers are in reducing production costs. This transition will be slow, and in the interim the future of Zephyrus is promising in regions like Hawaii, California, Texas, and New England. These regions have in common (1) a heavy dependency upon oil or gas for electric generation during the next twenty years, and (2) an abundance of windy sites a short distance from existing transmission lines or load centers.

Notes

1. The range of video interference is proportional to the size of the wind turbine blade, its rate of rotation, and the material from which it is made. The blades of 2-MW wind systems (200 feet in diameter) can interfere with reception as far as three miles away. Small systems (fewer than 25 kw) may interfere with reception within 100 feet (Senior and Sengupta, 1981).

2. For the particular windmill simulated, the ESI 54, a third-degree polynomial fits the speed-energy relationship better than a simple cubic equation.

3. Wind speeds have been characterized by a Rayleigh distribution. By knowing the mean wind speed of a site, the entire distribution can be estimated quite accurately. See Eldridge, 1980, Figures 111 and 112.

4. Cost estimates for small- and medium-size WECS were obtained from interviews with representatives of eight of the leading manufacturers at the American Wind Energy Association meetings in Amarillo, Texas, October 1982.

5. The author is grateful to Professor Edmond Cheng for providing the data. For his method, see Edmond Cheng and Benedict Wong, "Stochastic Simulation of Hourly Surface Wind in Hawaii," University of Hawaii, College of Engineering, April 1979.

6. Treating outages by derating results in an underestimate of the load on peaking units compared to a stochastic treatment. The downward bias is on the order of 10 to 15 percent. See Vardi and Avi-Itzhak (1981, chap. 2) for an explanation.

7. Three power pools in the United States and Canada are planning to generate nearly all electricity by coal or nuclear power by the end of the decade. By 1990, the utilities of the Mid-Continent Area Power Pool, Mid-America Interpool Network, and the East Central Area Reliability Coordination Agreement, from the Dakotas and Nebraska in the West, from Saskatchewan and Manitoba in the North, and to West Virginia and Kentucky in the East, expect to generate less than 5 percent of their energy from oil or gas by 1990 (NERC, 1982a).

8. Some of the risks of failure of individual WECS on a wind farm are not independent. Hurricanes, earthquakes, or fires can damage a group of windmills located in a small area. Furthermore, the failure of a transmission system (lines and transformers) from a wind farm to a grid can effectively disable all windmills. The risks of more common failures, such as the breaking of a blade or the wearing out of a bearing, are likely to be independent.

9. For a description of the statistical procedure for estimating the amplitude of cycles, see Peter Bloomfield, *Fourier Analysis of Time Series: An Introduction* (New York: John Wiley & Sons, 1976). The standard deviations and autocorrelations are estimated from residuals of a model with daily, semiannual, and annual cycles.

10. Weather data are generally autocorrelated. A heat wave or a cold spell refers to a sequence of days that are abnormally hot or cold. If it is especially hot today, you can predict with better than a flip of the coin that it will be especially hot tomorrow. Such predictable deviations from normal would be represented by a high positive autocorrelation coefficient, approaching +1.0. An autocorrelation of zero indicates that it is impossible to predict how the weather at 3 P.M. will deviate from normal if you know how the weather at 2 P.M. deviates from normal. For an example of English insolation sequences, see Brinkworth, 1977.

6
Central Power and the Impact of Photovoltaics on Electric Utilities

Stand-alone solar electric systems, serving facilities unconnected to utilities, are likely to be a minor factor in the future of highly developed economies. If a solar electric system were to meet a building's total energy demand, the inhabitants would have to choose between bearing extraordinarily high storage costs or experiencing a lifestyle of great energy austerity. More likely, solar electric technologies will be part of multifuel energy systems. While photovoltaics lend themselves to decentralized adoption, they can also be centralized in power stations. The modularity of photovoltaics enables 1,000 one-kilowatt systems to convert sunlight to electricity as efficiently as one 1,000-kilowatt system.

As a parallel to the previous chapter on wind-energy conversion systems, we examine several impacts of photovoltaic penetration on utilities, especially on their capacity and fuel requirements. These effects are examined for the same five representative utilities as before: Boston Edison, El Paso Electric, Florida Power and Light, Texas Utilities, and Hawaiian Electric. In addition, effects on two idealized coal-burning utilities are examined. El Paso and Hawaii enjoy the highest levels of insolation in the nation. Insolation in Fort Worth and Miami is representative of the semiarid Southwest and humid Southeast, respectively. Insolation in Boston is representative of the humid Northeast and Midwest. While the five sites differ in their load patterns and levels of insolation, the impact of photovoltaics on the respective utilities is strikingly similar in both direction and magnitude. For this reason, most of the results for oil- and gas-burning utilities will be illustrated for Boston and El Paso, which again represent the extremes. The summer-peaking idealized or "synthetic" utilities represent current conditions in the Midwest and future conditions elsewhere, when most utilities will have nearly eliminated gas and oil from their capacity mix and moved into coal, and possibly into nuclear plants.

The output of photovoltaic arrays is simulated from global horizontal insolation and temperature data for a "typical meteorological year." Compared to horizontal orientation, arrays installed at the same tilt as the latitude may receive, on clear days, 10 percent more insolation in the lower latitudes and 20 percent more in the higher latitudes. Mechanisms that track the sun can increase insolation by as much as 50 percent on a clear day, but these are unlikely to prove economical for decentralized installations. On cloudy days, orientation becomes less significant (Rapp, 1981, chap. 3).

Centralization versus Decentralization

The major effects of photovoltaics on utilities are likely to be fairly similar whether the technology is centralized at power stations or decentralized among end-users or load centers.

Centralized power systems have several disadvantages. First, and most important, is the energy lost in transmission and distribution lines between generator and consumer. Because of these line losses, as much as 1.1 kwh must be generated centrally to deliver 1 kwh to the end user. On-site photovoltaics would not suffer these frictionlike losses for electricity used on site; they would suffer these losses in any electricity sold back.[1] Second, while centralized stations must dedicate land to generating facilities, decentralized facilities can utilize vacant rooftops. Third, in the event of a blackout or catastrophic damage to central power stations or transmission lines, failures can cascade throughout the entire electrical network. The simultaneous failure of all decentralized systems is somewhat less likely.[2]

Decentralized utility-interactive electric power systems also may have disadvantages. They may require expensive modifications of the distribution end of the electric grid, particularly for two-way metering and for safeguarding the quality of electricity of other customers. These modifications include: (1) improved secondary isolation capability; (2) changes in distribution transformer capacity; (3) power-factor correction; (4) two-way voltage regulation; and (5) two-way metering.

Linemen face a potential safety hazard if broken power lines are inadvertently energized by decentralized photovoltaic systems. While it is possible to design inverters that generate AC power only in the presence of a utility-originated reference signal, the utility must guarantee that these devices will be properly maintained. A simple utility-controlled isolation switch can be mounted on a pole for easy disconnection. These switches cost about $20 per installation, or about 10 percent of no-solar, baseline distribution costs (Kammer, 1979).

Like other components of the electrical network, substation transformer capacity is normally geared to the peak flow of electricity in either direction. So long as the sellback is less than the baseline flow from the utility to the customer, photovoltaics can reduce capacity requirements of the substation transformers. If photovoltaic penetration is very high, however, there is a chance that inflows to the transformer will exceed baseline outflows. On balance, it is not clear whether decentralization provides an increase or a reduction in substation transformer capacity.

The quality of the electricity sold back to the utility may be inadequate. A cheap or faulty inverter (a device for converting the DC output of the array into common AC) can produce electricity with a low "power factor," or energy value. Utilities might require cogenerators to install a high-quality inverter, which is available for a marginal cost of about $300.

Voltages drop in the direction of power flow. Utilities install regulators to ensure that voltages are high enough at the end of the distribution lines. In periods of high feedback to the utility, voltages may be higher at the end of the line than at the distribution substation. This condition may necessitate the installation of two-way voltage regulators.

Metering requirements depend upon the rate structures established for sales and purchases by the utility. Under the current pattern of flat rates, a utility may permit the existing meter to move backward during sellback, or it may require a second one-way meter. Such simple meters cost only about $65. The institution of time-of-day pricing necessitates the installation of more complex meters, costing about $400. Installation of a two-way time-of-day meter may cost an additional $250.

If time-of-day rates are appropriately calculated to reflect marginal costs, a case can be made that the buying and selling prices should be nearly identical. Utilities would recapture distribution and overhead costs by a fixed fee. The acceptance of this argument would obviate the need for additional metering and permit the meter to operate in both directions.

Finally, decentralized systems may be more hazardous to the end user and more expensive to maintain and repair. Centralized systems may enjoy some scale economies in the balance of the system, but these appear to be small.

The Federal Energy Regulatory Commission (44 FR 12214, 25 February 1980) has ruled that utilities may neither favor nor discriminate against cogenerators vis-à-vis their other customers. This means that photovoltaic cogenerators will have to pay the marginal capital costs of making decentralized photovoltaic systems capable of interacting with utilities. The cost of isolation switches, adequate power-factor correction, and metering may total as much as $750, or three times current distribution costs per customer. These items add about 4 percent of the price of a 5-kw system anticipated in 1985.

Decentralized systems can save about 10 percent of central generation lost in transmission and distribution, but this advantage is lost if sellback is significant, which is the probable case. The advantages and disadvantages of centralization more or less balance, and at any rate they are probably of a small magnitude compared to the uncertainties surrounding the performance of photovoltaics.

How Utilities Select Their Capacity

It is useful to understand how utilities select their level and mix of capacity, and how they dispatch this capacity over the load cycle. The cost of generating electricity depends upon the cost of fuel, the generating equipment, and the labor to administer and maintain this equipment. In their investment decisions, utilities can select among several types of generators, with their own characteristic fuel, capital, and maintenance costs. Utilities try to choose the mix of plants that meets its anticipated load profile at least cost. While this freedom of choice has been restricted somewhat by the Power Plant and Industrial Fuel Use Act of 1978, these constraints can be ignored for the moment.

The ranking of capital costs for various generating plants is inverse to the ranking of fuel costs. The cost of building a gas-turbine plant "overnight" is only about $275/kw in 1980 dollars. On the average, oil-fired boilers cost about $600/kw, coal-fired generators about $800/kw, and light-water nuclear reactors about $1000/kw (*Table 6.1*). Interest payments during construction add more than 30 percent to the cost of a nuclear plant, 20 percent to the cost of a coal plant, and 15 percent to the

Table 6.1 Parameters for Economic Impact Analysis ($ 1980)

Utility costs	
"Overnight" capital costs: generation ($/kw)	
Nuclear	$1,000
Coal-fired	$800
Oil-/gas-fired boiler	$600
Oil-/gas turbine	$275
Capital costs: transmission ($/kw)	$200
Fuel costs/MBtu (1980)	
Nuclear	$1.00
Coal	$1.00-2.30
Oil-/gas-fired boiler	$6.00
Diesel oil turbine	$7.50
Operation & maintenance	0.2 cents/kwh
Annual capital charge rate	5-10%
PV system costs per kw	
Capital cost of system	
1980	$15,000
1985	$ 4,000
1990	$ 1,200
Operation & maintenance = 1% of capital costs	
Fuel-escalation rates (real)	2/3/4%
Discount rates (real)	5/8/10%

Source: Landsberg et al. (1979, Table 12-1, pp. 418-21); Schurr et al. (1979, chap. 9, pp. 272-73; Baughman, Joskow, and Kamat, 1979, Tables D.7, D.8; OECD, 1982, Table 7.12). Figures inflated to 1980 dollars and rounded.

cost of an oil or gas plant. In comparison, photovoltaic plants have practically been built overnight (PI, May 1983).

Capacity costs must be paid no matter how many hours a generator is used. On an annual basis, capacity cost depends upon the lifetime of the generator and the interest rate prevailing when the plant was built. Including taxes and insurance as well, real annual capital costs are about 5 to 10 percent (Stauffer, 1983).

For each plant type, the fixed annual capacity cost is represented graphically by the intercept, the variable energy cost by the slope. Each curve relates total generating cost to hours of operation. The overlay of the cost curves of the various plant types identifies the cheapest type of capacity to meet loads of varying duration (*Figure 6.1a*).

116 *Solar and Wind Energy*

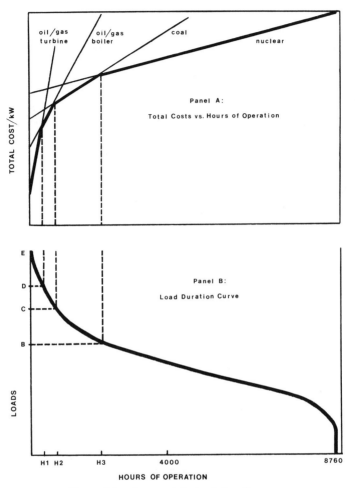

Figure 6.1 Costs and the load duration curve.

For a given number of hours of planned operation, the lowest curve identifies the cheapest plant type. For demands that last more than H3 hours, nuclear power is the cheapest. For demands that last between H2 and H3, coal power is the cheapest. Demands of more than H2 hours are called base loads, the irreducible minimum load on the system. For loads between H1 and H2 hours, oil-fired plants are the cheapest. Finally, for infrequent peak demands, which are expressed for less than H1 hours, gas-fired capacity is the most economical. Under the costs assumed here, H1 is about 300 hours, H2 about 700 hours, and H3 about 2,000 hours.

While there are interregional differences in capital and in most fuel costs on the order of 10 to 20 percent, the relative positions of the cost curves are fairly similar from one utility to the next.

The "load-duration curve" provides a graphic method for determining the amount of each capacity type a utility requires. This curve expresses in decreasing order the loads on a utility over the 8,760 hours of the year. The left-most point indicates the peak load, the hour of highest demand. The second point reflects the demand during the hour with the second-highest load, and so forth. The last and 8,760th point reflects the hour with the lowest load on the utility.

By juxtaposing the cost curves for four generator types with the load-duration curve, the base-, intermediate-, and peak-load periods are determined (*Figure 6.1b*). Capacity of OB is required to meet base loads that last for more than H3 hours. Capacity of BC is required to meet intermediate loads that last from H2 to H3 hours. Capacity CD is required to meet intermediate loads that last from H1 to H2. Finally, a capacity of DE is required to meet peak demands that last for less than H1 hours. Total capacity is OE. This capacity mix is the long-run optimum desired by the utility facing this hypothetical load-duration curve. The utility would select nuclear reactors for half of its capacity, coal for one-quarter, and oil or gas for one-quarter. El Paso Electric is likely to have a mix like this by 1990.

Because generating plants undergo routine maintenance and accidental breakdowns, utilities must keep spare capacity in reserve. One way of calculating the required reserve margin is to derate capacity by the availability of the unit. If a coal plant is available only 80 percent of the time because of forced and scheduled outages, then in order to have 1,000 MW of coal-fired capacity available, about 1,250 MW of capacity (1,000/.80) must be installed.

Because of the longevity of generating equipment, few utilities have achieved their optimum capacity mix. Most have far more oil or gas capacity than desired and far less coal or nuclear capacity than desired. Not surprisingly, utilities have begun a concerted shift back to coal (NERC, 1983). These market forces have received a fillip from the Power Plant and Industrial Fuel Use Act of 1978, which encourages conversion to coal and prohibits the use of oil and gas in base-load plants after 1990. Fuel economics render this act superfluous.

The reduction in peak transmission capacity is proportional to the reduction in peak generating capacity. Palz (1978) estimates transmission costs at $200/kw. Averages are misleading, however, because transmission costs depend upon the length of the lines, the population density of the service area, and the mix of industrial and residential

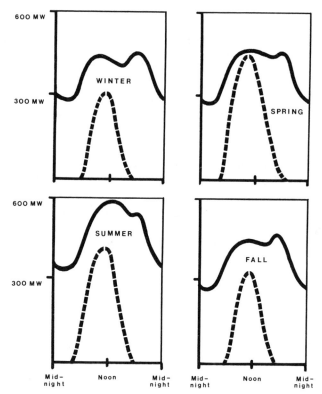

Figure 6.2   Average daily profiles, El Paso: Loads and array output. (Loads indicated by solid lines; windmill output, by dotted line).

customers (Baughman, Joskow, and Kamat, 1979, chap. 3). The value of transmission-cost savings from photovoltaics will be utility-specific.

Capacity Savings: The Impact on the Annual Load Profile

As is the case with wind, the daily load profile is not well synchronized with insolation. The rhythmic rising and setting of the sun generates a familiar bell curve of photovoltaic output (*Figures 6.2 and 6.3*). Current patterns of electric use hardly conform to the sun's daily rhythm. First, the daily load pattern rarely peaks at noon. Second, except in summer, seasonal loads peak when insolation is low. Third, loads are substantial at night, when insolation is nonexistent. These discrepancies between cycles of load and insolation suggest that photovoltaics are unlikely to shave the peak loads on utilities. Consequently, solar photovoltaics appear to contribute most to the reduction of daily shoulder loads.

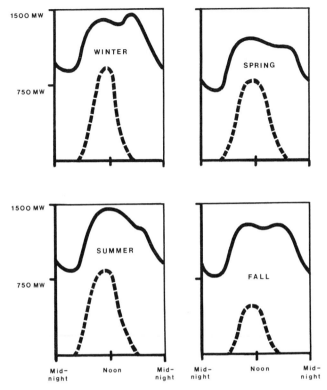

Figure 6.3 Average daily profiles, Boston: Loads and array output. (Loads indicated by solid lines; windmill output, by dotted line.)

Insolation is also low during the winter, when many utilities experience short spikes in demand. While the heaviest loads on Florida Power and Light occur during the summer air-conditioning season, the actual peak occurs on a winter morning, when electric-resistance heating combats a cold snap. Boston Edison's loads are about as high during the winter, when insolation is low, as during the summer. Thus seasonal patterns of loads and insolation hinder peak shaving.

Despite differences in load profiles among the utilities, the impact of photovoltaic penetration on the peaks are fairly similar (*Table 6.2*). One megawatt of photovoltaics reduces peak load by less than one megawatt. In Hawaii, PV penetration has virtually no impact on peak load. In El Paso, a bit more than 1 MW of photovoltaics is required to displace 1 MW of conventional capacity at the 3 percent penetration, but 2 MW of photovoltaics are required in Boston, 6 MW in Fort Worth, and 9 MW in Miami.

Table 6.2 Peak Load, Average Load, and Load Factor at Different Levels of Photovoltaic Penetration

% PV penetration	Peak load	Average load	Load factor
Hawaii			
0	917	710	.774
3	917	703	.767
10	917	688	.750
20	917	666	.726
30	917	644	.703
Boston			
0	2100	1203	.573
3	2061	1194	.579
10	2051	1172	.572
20	2039	1142	.560
30	2026	1111	.549
Fort Worth			
0	11202	6183	.552
3	11148	6113	.548
10	11023	5951	.540
20	10935	5719	.523
30	10893	5486	.504
El Paso			
0	688	416	.605
3	669	411	.613
10	677	399	.589
20	666	382	.574
30	659	365	.554
Miami			
0	8860	5169	.583
3	8832	5120	.580
10	8766	5006	.571
20	8671	4843	.558
30	8577	4680	.546

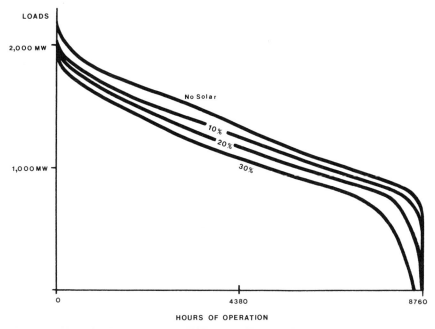

Figure 6.4 Load duration curves, various levels of solar penetration: Boston Edison.

Second, successive increments in photovoltaic capacity generally result in diminishing reductions in peak load. At the 30 percent penetration, 7 MW of photovoltaics are required to displace 1 MW of conventional capacity in El Paso. To effect the same displacement, 8 MW are required in Boston, 11 MW in Forth Worth, and 9 MW in Miami.

The effect of photovoltaics on peak load is similar in magnitude to the impact of wind-energy conversion systems. In Hawaii, neither has much impact on peak load. In windy Boston, WECS have somewhat greater impact on peak load than photovoltaics. In sunny El Paso, Fort Worth, and Miami, photovoltaics have greater impact than WECS. These minor differences aside, utilities that planned their capacity requirements on the basis of peak load plus a reserve margin would enjoy rather minor benefits from either photovoltaic or WECS penetration.

Third, the changes in load factors indicate that photovoltaic penetration results in reductions of average load faster than peak load. This means that photovoltaic penetration will change the optimal mix of capacity.

The load-duration curves for Boston illustrate clearly where the reductions in load occur (*Figure 6.4*). The heavy line indicates the

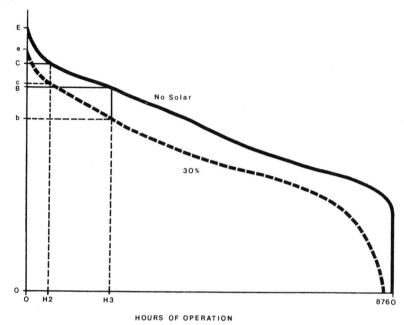

Figure 6.5 Effects of photovoltaic penetration on desired mix of generating capacity: Boston Edison.

current, no-solar load-duration curve. The lower curves indicate the effects of photovoltaic capacity in increments of 10 percent of peak load. Since all the curves converge on the vertical axis at the same point, only the right-hand portions of the curves are easily distinguished from each other.

Ten percent penetration of photovoltaic systems lowers the middle portion of the load-duration curve, which is indicative of reduced shoulder-load capacity requirements. Greater penetration reduces base loads disproportionately. As penetration exceeds 20 percent, photovoltaic generation exceeds the load for a few hours of the year. This surplus would be wasted if it could not be stored or sold to another utility.

The implications of solar penetration for utility-capacity planning can now be drawn. The generating-cost curves of Figure 6.1a indicate the point H3, where there is a switch from nuclear to coal, and point H2, where there is a switch from coal to gas. These points are based upon relative costs and are unaffected by changes in the load-duration curves. These points are indicated on graphs that show the load-duration curves of the no-solar baseline and 30 percent solar penetration in Boston (*Figure 6.5*).

At 30 percent photovoltaic penetration, total capacity requirements drop from OE to Oe. Base-load capacity requirements drop even more, from OB to Ob. Even though the curve falls most substantially in the shoulder loads, shoulder-load capacity requirements increase from CB to cb, an apparent paradox. Finally, required peaking capacity increases slightly from EC to ec, another apparent paradox.

This analysis indicates that increased penetration of solar photovoltaics results in a small decline in generating-capacity requirements, but it results in a major shift in its composition. If there were no legal or political constraints on choice of plant, photovoltaic penetration would reduce the desired base-load capacity (coal and nuclear) and increase the desired intermediate and peaking capacity (oil and gas).

Two implications of this analysis should be clarified. First, the twist toward oil and gas apppears to be contrary to the stated aims of national energy policy, which prohibits the expansion of gas and oil capacity. This inconsistency is more apparent than real. While gas and oil capacity might increase, the utilization of these fuels is likely to decrease as a result of solar penetration.

Second, while photovoltaic penetration may reduce the desired amount of nuclear capacity, the amount of nuclear capacity desired on economic grounds probably exceeds the amount likely to be installed in the near future. Even by 1990, many utilities will have relatively too much gas and oil capacity. If photovoltaics permitted utilities to phase out existing capacity, simulations noted below indicate that phasing out oil and gas would reduce costs more than phasing out nuclear or coal plants.

Capacity Savings: The Impact on Reliability

The load-duration curve is useful in demonstrating the effect of photovoltaics on the desired capacity mix. As noted in the analysis of wind farms, utilities are increasingly using the loss-of-load probability as a criterion for planning total capacity requirements.

In replicating the analysis of wind farms, it is interesting to focus some attention on the effects of renewables on the maintenance of conventional capacity. The routine maintenance of conventional generators is scheduled to levelize the loss-of-load probability throughout the year. Maintenance may take as long as eight weeks, so the best periods for scheduled maintenance are those with low weekly peak loads. In practice, the criterion of levelized loss-of-load probability results in scheduling maintenance for spring and fall, when loads are relatively low.

Weekly peaks at two levels of solar penetration are indicated in Figure 6.6. Even if photovoltaic capacity is as high as 30 percent of peak load,

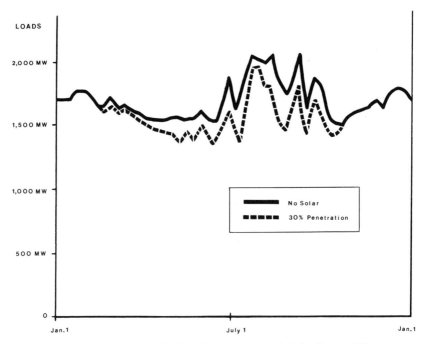

Figure 6.6 Weekly peak load and maintenance schedule: Boston Edison.

the weekly peaks are reduced only slightly during the spring and fall, when maintenance is normally scheduled. Few of the reductions in the late winter last long enough to allow any major additional maintenance efforts. The most prolonged reduction occurs during the summer peak, when maintenance is never scheduled and capacity is reserved to meet forced outages and unanticipated spikes in demand. Solar penetration will thus provide little gain in flexibility to the utility in scheduling maintenance.

If photovoltaic systems are decentralized among end-users, the scheduling of their maintenance will occur in an uncoordinated manner. Involving merely washing and cleaning the arrays, routine maintenance is unlikely to remove a high percentage of capacity at any time. Even if all households cleaned their arrays on the same spring weekend, output would hardly diminish. The balance of the photovoltaic system may require periodic inspection, maintenance, and replacement at the same rate as conventional generating capacity.

Because of their modular nature, photovoltaics are likely to experience forced outages randomly throughout the year. A given proportion of capacity is likely to be down at any one time. What that proportion is cannot be determined yet. Several studies assume a 1 percent forced outage rate for these systems (General Electric, 1978, vol. 2, E-43 to 49; Finger, 1980). Field tests eventually may provide more solid information about these outage rates.

Results

As before, loss-of-load probabilities of Boston Edison and El Paso Electric are calculated for the no-solar baseline and for four levels of photovoltaic capacity using Program LOLP. They are also computed for several levels of utility capacity in order to determine the savings in capacity that solar penetration will permit without increasing the loss-of-load probability. Capacity reductions fall upon gas units in El Paso and upon oil units in Boston.

The results are plotted in Figure 6.7. The horizontal axis marks conventional capacity, in intervals of 21 MW for El Paso Electric and 56 MW for Boston Edison. These intervals represent about 2 percent of projected capacity. The vertical axis marks the loss-of-load probabilities, with the baseline highlighted by a dotted horizontal line. Each curve represents a given level of photovoltaic capacity. The necessary conventional capacity is indicated by the intersection of each curve with the dotted baseline. For example, El Paso's 10 percent penetration curve intersects the baseline at 1,012 MW. This means that the target loss-of-load probability can be achieved by El Paso Electric with a 4 percent reduction in conventional capacity if photovoltaic capacity is 10 percent of peak load.

The capacity reductions indicated by this measure of reliability are shown in Table 6.3. At higher penetrations, the indicated reductions tend to be larger than under the reserve-margin approach. Moreover, the indicated capacity reductions increase more rapidly under the reliability approach than under the reserve-margin approach. At 30 percent penetration, the indicated capacity reduction is twice as high under the reliability approach.

If photovoltaics replace $600/kw oil plants in Boston, the value of the capacity credit ranges from $360 per kilowatt of photovoltaics at low penetrations to $180 per kilowatt at high penetrations. If $275/kw gas boilers are replaced in El Paso, the savings range from $140 to $120.

The Impact on the Volatility of Hourly Loads

Like wind penetration, increased photovoltaic penetration reduces the amplitude of daily cycles. In all sites the reduction in amplitude is substantial. Photovoltaic penetration reduces the amplitude relative to

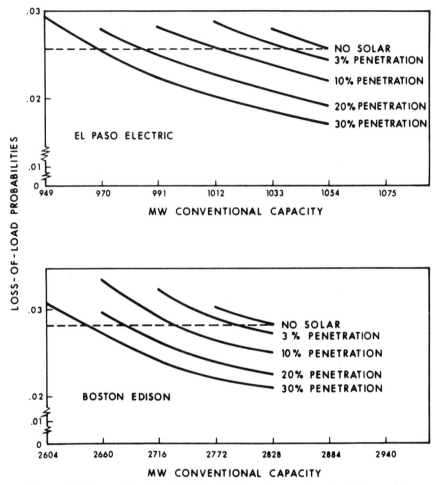

Figure 6.7 Loss-of-load probabilities: Effects of conventional and photovoltaic capacity, El Paso and Boston.

the mean load, while wind penetration leaves this ratio unchanged (*Table 6.4*).

In Hawaii, Boston, and El Paso, photovoltaic penetration slightly raises the standard deviation of hourly loads around these cycles, but the effect is far less than that of wind penetration. Finally, photovoltaic penetration at the three sites leaves the pattern of autocorrelations virtually unchanged. These findings suggest that neither solar nor wind penetration is likely to make the dispatcher's job more difficult.

Table 6.3 Peak-Load Reduction, Capacity Reduction, Load Factor: Various Penetrations of Photovoltaics

MW PV capacity	% of peak load	Load factor	Peak-load reduction	Peak-capacity reduction (res. marg. = .15)	Peak-capacity reduction (LOLP constant)
El Paso					
25	3.6	.613	18.6	21.4	13
75	10.9	.604	27.6	31.7	42
150	21.8	.585	35.0	40.3	67
200	29.1	.565	40.5	46.6	84
Boston					
60	2.9	.579	38.6	44.4	36
200	9.5	.572	49.3	56.7	92
400	19.0	.560	61.5	70.7	142
600	28.5	.549	73.8	84.9	182

The Value of Fuel Savings

In Chapter 4, the net economic benefits of residential photovoltaics were calculated on the basis of individual systems. A superior method of calculating benefits is the production-costing approach, which was applied to wind farms. The avoided-cost calculation is independent of the particular and somewhat hypothetical rate structure that a utility might impose in the future.

Fuel Savings

Boston and El Paso exemplify the range of regional differences in output in the continental United States (*Table 6.5*). Each kilowatt of photovoltaics in El Paso generates about 2,100 kwh per year. This is more than twice the output of windmills at the same site. Each kilowatt of photovoltaics in Boston generates only about 1,250 kwh a year. This figure is about two-thirds of the wind output at the same site.

There are striking similarities between the effects of wind penetration and photovoltaic penetration on the utilities in both sites. First, photovoltaic penetration saves a disproportion of the expensive fuels, oil or gas. Second, as penetration increases, the share of expensive fuels saved becomes smaller. Photovoltaic output, however, appears to be better synchronized with the use of peak fuels than windmill output. At

Table 6.4 Fluctuations in Utility Loads under Various Photovoltaic Penetrations

% PV penetration	Average load MW	Daily amplitude MW	Standard deviation MW	Autocorrelation		
				1-hour	2-hour	3-hour
Hawaii						
0	710	185	71	.87	.62	.31
3	703	177	71	.87	.62	.30
10	688	161	73	.87	.61	.29
20	666	141	78	.86	.59	.27
30	644	128	86	.85	.57	.24
Boston						
0	1203	293	198	.95	.84	.72
3	1193	282	198	.95	.85	.72
10	1172	260	199	.95	.85	.73
20	1142	232	205	.95	.85	.72
30	1111	212	215	.95	.84	.71
El Paso						
0	416	97	54	.92	.76	.58
3	411	91	53	.92	.76	.58
10	399	78	54	.92	.76	.58
20	382	66	57	.91	.74	.53
30	366	63	61	.90	.71	.47

30 percent windmill capacity, Boston's oil-fired plants bear about 70 percent of the reduction in conventional fuel. At the same level of photovoltaic capacity, such plants bear about 80 percent of the fuel reduction. In El Paso, at 30 percent wind capacity, less than 50 percent of the reduction in conventional generation is borne by gas-fired generators. At a similar photovoltaic capacity, over 60 percent of the reduction is borne by gas.

The Public-Policy Perspective

The savings in conventional fuel provide a first approximation of the value of a kilowatt-hour of photoelectricity (*Table 6.6*). The values are lowest in Fort Worth, where there is a disproportionate savings in inexpensive coal, the highest in Hawaii, where all savings are in oil. In Boston, the value of the fuel mix falls from 5.7 cents/kwh to 5.2 cents/kwh with increased photovoltaic capacity. These values are about 40 mills/kwh more than those produced by windmills. In El Paso, the

Table 6.5 Kilowatt-Hours of Electric Generation Displaced per Kilowatt of PV Capacity

	PV Penetration as % of Peak Load			
	3	10	20	30
Honolulu				
All oil	2073	2073	2073	2073
Boston				
Nuclear	139	165	213	255
Oil	1137	1102	1053	1011
total	1276	1276	1276	1276
Fort Worth				
Nuclear	0	0	0	0
Coal	699	803	991	1146
Gas	1117	1012	825	670
total	1816	1816	1816	1816
El Paso				
Nuclear	82	142	236	363
Coal	255	257	330	430
Gas	1777	1715	1548	1321
total	2114	2114	2114	2114
Miami				
Nuclear	0	0	0	0
Coal	22	40	86	154
Oil	1592	1574	1527	1460
total	1614	1614	1614	1614

value of the fuel mix is 5.4 cents/kwh at low penetrations and 4.5 cents/kwh at high penetrations. These are 40 to 60 mills/kwh greater than those produced by windmills. The higher value of savings from photovoltaics is traceable to the superior synchrony between load and insolation cycles.

In 1980, the levelized cost of photovoltaic systems was about ten times the current fuel cost in Hawaii and El Paso. In Boston and Fort Worth, the factor was about twenty times. For 1985 installations, costs are three to six times greater than levelized benefits under the worst-case scenario and about two times greater under the best-case scenario. By 1990, benefits exceed costs in all sites under the best-case scenario and in Hawaii and in El Paso under the worst case.

The discounted present value of the fuels and operating costs saved by

Table 6.6 Levelized Cost and Savings of PV-Generated Electricity (cents/kwh)

	Honolulu	Boston	Fort Worth	El Paso	Miami
Social cost					
1980 installation					
5% discount	53.2	106.9	74.5	64.0	83.8
10% discount	75.2	132.5	92.3	79.3	103.9
1985 installation					
5% discount	14.2	28.5	19.8	17.0	22.3
10% discount	17.6	35.3	26.8	21.1	27.7
1990 installation					
5% discount	4.2	8.5	5.9	5.1	6.7
10% discount	6.0	12.0	8.4	7.2	9.4
1980 value of fuel					
3% penetration	6.3	5.7	4.1	5.4	5.9
30% penetration	6.1	5.3	2.8	4.5	5.6
Levelized value of fuel (3% penetration)					
(2% fuel/10% discount)					
1985 installation	8.0	7.1	5.1	6.9	7.5
1990 installation	8.7	7.7	5.6	7.4	8.1
(4% fuel/5% discount)					
1985 installation	11.6	10.4	7.4	10.0	10.9
1990 installation	13.6	12.2	8.7	11.7	12.8

photovoltaics is computed at the same three discount and fuel-escalation rates as before for systems installed in 1980, 1985, and 1990. At first, calculations based upon social-replacement costs exclude considerations of externalities and taxes. The most illustrative scenarios reflect the best case for photovoltaics, the worst case, and the most likely case (*Table 6.7*).

1. As with windmills, fuel is responsible for the bulk of the savings. Savings in operations and maintenance are negligible, less than 3 percent of fuel savings.
2. As expected on the basis of both photovoltaic output and fuel costs, the value of fuels saved is higher in El Paso than in Boston. Savings in Fort Worth and Miami are intermediate between these two.
3. The value of fuel savings per kilowatt of photovoltaics diminishes with penetration. As with windmills, increased photovoltaic penetration saves increasingly cheaper base and intermediate fuels.

Table 6.7 Present Worth of Fuel Savings per Kilowatt of Photovoltaic Capacity: Alternative Scenarios, 1980-90, El Paso and Boston

	PV Penetration as % of Peak Load			
	3	10	20	30
El Paso				
Worst case: 2% fuel/10% discount				
1980 installation	1215	1121	997	926
1985 installation	1342	1239	1103	1024
1990 installation	1484	1370	1218	1131
M&O	34	34	33	32
Capacity	294	203	117	79
Most likely: 3% fuel/8% discount				
1980 installation	1563	1443	1283	1191
1985 installation	1817	1677	1492	1384
1990 installation	2110	1948	1733	1608
M&O	40	39	39	37
Capacity	294	203	117	79
Best case: 4% fuel/5% discount				
1980 installation	2287	2111	1878	1743
1985 installation	2795	2581	2295	2131
1990 installation	3411	3149	2801	2600
M&O	51	51	50	48
Capacity	294	203	117	79
Boston				
Worst case: 2% fuel/10% discount				
1980 installation	803	790	780	758
1985 installation	887	872	862	838
1990 installation	981	965	952	926
M&O	20	20	20	20
Capacity	306	178	110	79
Most likely: 3% fuel/8% discount				
1980 installation	1033	1016	1003	976
1985 installation	1201	1187	1182	1135
1990 installation	1395	1372	1352	1318
M&O	23	23	23	23
Capacity	306	178	110	79
Best case: 4% fuel/5% discount				
1980 installation	1512	1487	1468	1428
1985 installation	1840	1817	1794	1745
1990 installation	2255	2218	2190	2131
M&O	30	30	29	29
Capacity	306	178	110	79

132 *Solar and Wind Energy*

4. Capacity savings comprise a significant share of total savings at penetrations below 10 percent. At higher penetrations, capacity savings are only about one-tenth of fuel savings. The relative significance of capacity savings diminishes over time because of fuel escalation.

The total value of savings is far below $15,000/kw, the 1980 cost of residential photovoltaics. If system costs follow the projections in Chapter 3 and if fuel prices continue to escalate faster than inflation, then the picture improves over time. By 1985, system costs are projected at $4,000/kw, still above the present value of fuel benefits in both sites under all reasonable scenarios. By 1990, however, projected costs fall to about $1,200, which makes photovoltaics potentially interesting to utilities in both cities. If central power stations can be built with a substantially lower balance-of-systems cost than decentralized arrays, the picture may prove to be attractive even earlier.[3]

Comparing Perspectives

The effects of adding external costs and tax considerations are easily seen from the levelized costs and benefits of photovoltaic systems. The benefits include the conventional fuel, maintenance, and capacity saved by the utility.

In 1985, the levelized value of system costs still exceeds the levelized benefits, even when a 25 percent surcharge on fuels is included. The surcharge would have to exceed 50 percent to justify photovoltaics in El Paso even under the best-case scenario. By 1990, however, a 25 percent surcharge is effective enough to justify photovoltaics in Boston under the worst-case scenario.

The tax laws in force during the early 1980s make investments in renewable energy most favorable for homeowners and least favorable for utilities. This means that homeowners can generate electricity for themselves more cheaply than can a third party or a utility (*Table 6.8*). This difference can be traced almost entirely to differences in tax credits rather than to depreciation rules.

In 1985, tax incentives are insufficient to make photovoltaics profitable from any of the three perspectives. By 1990, all three investors will find photovoltaics profitable in El Paso. Only utilities would not find photovoltaics profitable in Boston in 1990.

Table 6.8 Levelized Costs and Benefits of Photovoltaics: Four Investment Perspectives, 1990

	Public Policy			Utility*	Third-Party**	
	Benefits		Social cost	Required revenues	Required revenues	Homeowner cost
	0%	25%				
Boston						
2% fuel/10% discount	10.5	12.8	12.0	14.6	10.1	8.2
4% fuel/5% discount	14.4	17.5	8.5	9.5	6.7	6.1
El Paso						
2% fuel/10% discount	9.0	10.9	7.2	8.7	6.0	4.9
4% fuel/5% discount	12.8	16.2	5.1	5.7	4.0	3.6

*Utility faces 40 percent marginal tax rate, 10 percent tax credit, and 10-year straight-line depreciation.

**Third-party investor faces 50 percent marginal tax rate, 25 percent tax credit, 5-year accelerated cost recovery.

Renewable Energy as Insurance

The profitability of investments in wind and photovoltaics has been shown to be highly sensitive to fuel-escalation rates. Because of the fuel-adjustment clause, electric ratepayers bear all of the risks of fuel escalation, while the utility bears none. In other words, renewable energy systems serve as insurance from the viewpoint of the homeowner, but not from that of the utility. The institution of the fuel-adjustment clause leads utilities to undervalue renewable energy as insurance.

How does the penetration of wind and photovoltaics affect the risks of escalating electricity rates? The electricity bills in the five sites are examined under four conditions in 1990: (1) no renewables; (2) 20 percent photovoltaic penetration; (3) 20 percent wind penetration; and (4) 10 percent photovoltaic/10 percent wind penetration. Four sets of contingencies are examined:

- fuel escalation: 1/2/3/4/5 percent
- performance as proportion of specifications: .9/1.0/1.1
- maintenance as proportion of capital costs: .01/.02/.03 (WECS)/ .005,.01,.015 (PV)
- longevity of system: 15/20/25 years

Wind systems installed in 1990 are costed at $1,500/kw and photovoltaic systems at $1,200. Annual costs and benefits in all scenarios are discounted at 8 percent. A capacity credit, if any, is added to the present value of fuel savings.

In all five sites, the penetration of WECS and photovoltaics reduces the standard deviation in potential utility-generating costs by 12 to 18 percent. The expected values of both sources of renewable energy were positive in Hawaii and Boston, but the expected value of WECS penetration was negative in the other three sites. In Hawaii and Boston, the 20-percent-WECS "portfolio" provides the largest reductions in both the expected value and variation (risks) of fuel costs. In El Paso and Miami, the 20-percent-photovoltaic portfolio provides the largest reductions in both the expected value and variation in fuel costs. In these four sites, the portfolios indicated provided both the highest return and the lowest risk.

The Fort Worth case is somewhat more complicated. The 20-percent-WECS portfolio has an expected generation cost about 5 percent higher than the no-solar baseline, but the variation in fuel costs are 13 percent less. The 20-percent-photovoltaic portfolio reduces expected generation cost by 8 percent and reduces the risk by 18 percent, which makes this clearly superior to the previous portfolio. The mixed portfolio with 10 percent WECS/10 percent photovoltaics reduces expected generating costs by 8 percent, but it reduces risk by only 16 percent. The choice between these last two portfolios depends upon the public's relative weighting of risk and return.

From 70 to 90 percent of the variance in the net savings from these renewable systems is traced to uncertainty in fuel escalation and the rest to the remaining technological uncertainties. Under the current institution of the fuel-adjustment clause, the utility bears all of the technological risks of failures of renewable systems, but it obtains none of the gains in fuel savings, which are passed on to the ratepayers. Although relatively small, these technological uncertainties increase the riskiness of the utility's earnings, but they provide the utility with no offsetting gain.

Photovoltaics and the Transition to Coal

So far we have assumed that as the cost of photovoltaics or wind-energy conversion systems fall and the cost of conventional fuels rise, renewables will become increasingly attractive. This scenario ignores the slow but steady pace of fuel switching that has accompanied the rising costs of oil and gas. The impressive fuel savings resulting from photovoltaic penetration in the service area of oil- and gas-fired utilities may be atypical in the future. While many utilities may have considerable oil-

and gas-fired capacity by the end of the 1980s, this is only a transitory condition. In the long run, electric utilities are likely to reach a new equilibrium, in which nearly all of their capacity is comprised of coal-fired, and possibly nuclear, power plants. While synfuels may replace natural gas or oil as boiler fuels, their cost is unlikely to make coal less attractive. How are photovoltaics likely to fare in a coal-fired future?

To answer this question, we examine two "synthetic" utilities characterized by the Electric Power Research Institute.[4] These two representative utilities are ascribed the capacity mix that will be typical in the Midwest by 1990: 20 percent nuclear, 72 percent coal, and 8 percent oil turbine. Both utilities are summer-peaking, but the northern utility has a relatively higher winter load than the southern utility. Boston provides the isolation data for the northern utility; and El Paso for the southern utility. The 1980 cost of coal for both utilities is set at $2/MBtu, which is typical of prices in eastern sites but about twice as high as in western sites. Nevertheless, these assumptions provide an upper bound to the fuel savings offered by photovoltaics.

Not surprisingly, the value of fuel savings from photovoltaic electricity is dominated by the cost of coal. The value of savings in oil or uranium is trivial at all penetrations. In both utility areas, the 1980 value of fuel savings is less than 2.1 cents/kwh and decline slightly with increased solar penetration. If coal were priced at $1/MBtu, the value of fuel savings would be halved. Even allowing for a high rate of fuel escalation, the levelized value of these savings is less than 3 cents/kwh.

Because of greater insolation, both fuel savings and capacity savings are greater in the southern area than in the northern area. The ratios of both fuel savings and capacity savings between the two areas are two to one (*Table 6.9*).

The heavy capital costs and low fuel costs of coal-fired generators result in significant capital savings, relative to the results for oil- or gas-fired utilities. Depending upon the scenario, capital can contribute as much as half of the savings. Obviously, the higher the fuel-escalation rate, the lower the capital share in savings.

Photovoltaics would appear to be fairly uncompetitive for a northern coal-fired utility. By 1990, the projected system cost of $1,200/kwh exceeds estimated fuel savings, even under the best-case fuel-escalation scenario. At a low penetration, the sum of fuel, maintenance, and capital costs are less than the projected cost of photovoltaic plants built in 1990. In parts of the North where coal was priced at $1/MBtu in 1980, photovoltaics would be far from competitive.

For the southern coal-fired utility, photovoltaics are more competitive. By 1990, the value of fuel, maintenance, and capacity savings are slightly below the $1,200/kw system cost under the worst-case fuel-escalation

Table 6.9 **Present Worth of Savings ($/kw) of Photovoltaic Capacity: Alternative Scenarios, Synthetic Utilities**

	PV Penetration as % of Peak Load			
	3	10	20	30
Northern utility				
Worst case: 2% fuel/10% discount				
1980 installation	251	247	245	244
1985 installation	283	279	277	276
1990 installation	307	302	300	298
M&O	21	21	21	21
Capacity	279	167	131	107
Most likely: 3% fuel/8% discount				
1980 installation	323	318	316	314
1985 installation	387	380	378	376
1990 installation	436	429	426	424
M&O	24	24	24	24
Capacity	279	167	131	107
Best case: 4% fuel/5% discount				
1980 installation	473	465	462	460
1985 installation	601	591	587	585
1990 installation	705	694	689	686
M&O	31	31	31	31
Capacity	279	167	131	107
Southern utility				
Worst case: 2% fuel/10% discount				
1980 installation	422	414	410	407
1985 installation	475	467	462	459
1990 installation	515	506	501	497
M&O	35	35	35	35
Capacity	660	415	259	186
Most likely: 3% fuel/8% discount				
1980 installation	542	533	527	524
1985 installation	649	638	631	627
1990 installation	732	719	712	707
M&O	41	41	41	41
Capacity	660	415	259	186
Best case: 4% fuel/5% discount				
1980 installation	793	780	772	767
1985 installation	1009	991	981	975
1990 installation	1184	1163	1151	1144
M&O	52	52	52	52
Capacity	660	415	259	186

scenario at a low penetration. Under the most likely scenario, photovoltaics appear competitive at penetrations of nearly 10 percent by 1990. Under the optimistic scenario, they appear competitive at 30 percent penetration.

For portions of the southern tier where coal was priced at $1/MBtu, the competitive prospects for photovoltaics are far less, but they are not hopeless. Under the most likely scenario, a halving of fuel savings still leaves photovoltaics competitive at 3 percent penetration. Under the most optimistic scenario, a halving of fuel savings leaves photovoltaics competitive at a 10 percent penetration.

Conclusions

The numerical examples here assume that photovoltaics are installed in a central power station or sun farm. Instead, if the arrays were decentralized, the general conclusions would remain quite the same. In a decentralized mode, photovoltaics would provide some savings in transmission capacity and some reductions in line losses between the power station and the end-user. Together, these might enhance the capital and fuel benefits by about 10 percent. Balanced against these savings are costs of making the arrays utility-interactive, as much as $750 per 5-kw system. These costs of utility interaction would be substantially less if the utility simply permitted the meter to reverse under the sellback mode. If not, these costs might add about 5 percent to the cost of a 5-kw system in 1985.

Minor differences between centralized and decentralized locations aside, photovoltaic penetration of up to about 30 percent of peak load can have several positive effects on utilities that are now heavily dependent upon gas and oil.

First, even though less than half the capacity of the utilities considered will be comprised of gas or oil boilers by 1990, most of the savings from solar penetration will be in costly gas or oil. Since the value of gas and oil will be at least three times as great as that of coal, savings in gas and oil generation comprise the majority of all utility savings. Fuel savings dwarf savings in other operating costs.

Second, solar penetration reduces the peak loads on utilities slightly. In the absence of electric storage, the ability of photovoltaics to shave the peak further is limited because peak loads occur in the early evening during the summer and in the early morning during the winter. Peaks may flatten as utilities adopt load-management practices. If substantial photovoltaic penetration were to occur, utilities could save both capacity and fuel by shifting loads toward solar noon.

Third, photovoltaic penetration appears to have a small but positive impact on the loss-of-load probability of the utility. Together, the reductions in peak load and in LOLP permit the utility small capacity savings. For oil- and gas-fired utilities, the value of these capacity savings are about one-tenth the value of fuel savings.

Fourth, the effect of solar penetration on hourly load fluctuations of the utility is more complex. Photovoltaic penetration reduces the amplitude of systematic fluctuations in daily load cycles, but it increases the randomness of hour-to-hour variations in load slightly. All in all, the predictability of loads several hours in advance does not change much over a wide range of photovoltaic penetrations. As a result, the job of the dispatcher does not become more difficult.

In 1980 the social costs of photovoltaic systems were about seven to ten times greater than the value of fuel and capacity savings in El Paso and about ten to twenty times greater than the values in Boston. Even the 40 percent tax credit could not overcome this differential. Clearly, major cost decreases are required before photovoltaics can become attractive economically. If costs of photovoltaic systems follow the projections in Chapter 3, regions whose utilities are heavily dependent upon gas or oil will find it profitable by 1990 to encourage the installation of photovoltaics in central power plants, on sun farms, or on end-user premises. If cost reductions are realized, utilities in the southern swath between California and Florida and in the Northeast will comprise a major market for photovoltaics.

When other adjustments to rising oil and gas prices are considered, the bright prospects for photovoltaics become cloudier. The rising oil and gas prices, which make photovoltaics increasingly attractive, are also giving rise to a slow but steady shift of generating capacity toward coal. Several utilities in the Midwest may have completed the transition to coal by 1990, but for most the transition will take longer. The expanding role of coal in the utility industry's generation mix will offset the effects of rising oil and gas prices on the profitability of photovoltaics.

The economic impacts of photovoltaics in hypothetical coal-fired utilities in the northern and southern insolation belts of the United States are far less favorable than the impacts on oil- and gas-fired utilities. While the value of capacity saved by photovoltaics is higher for coal-fired utilities, the value of the fuel is far less. On balance, it appears that photovoltaics will not be economical for northern coal-fired utilities by 1990. The picture is somewhat more favorable for photovoltaics in portions of the southern insolation belt, where coal is relatively expensive.

The successful transition of utilities from an oil and gas base to a coal base can stifle the potential penetration of photovoltaics. This is particularly true in the northern regions, where insolation is relatively poor.

The completion of this transition is questionable, however, as issues of carbon dioxide buildup, acid precipitation, and other environmental and health impacts become more prominent. Another round of environmental legislation may raise the costs of coal generation further, rendering an all-coal future less attractive. The environmental advantages of photovoltaics may redress the balance.

Notes

1. A 10-kw residential photovoltaic system analyzed in Chapter 4 sells back about 42 percent of its output in New York City and 47 percent in Fort Worth. These high sellback proportions practically eliminate any savings in line losses from reductions in load on the utility (see Note 4 in Chapter 4).

2. A failure of one or two components can black out a power plant and even an entire power network. A single component failure also can black out a single photovoltaic system or even several systems sharing a common distribution network, but not the entire regional power network. All decentralized photovoltaic systems can be damaged by the "electromagnetic pulse" from an atomic explosion. In this eventuality, centralized and decentralized systems would appear to share a common vulnerability. For a discussion of vulnerabilities of power stations and grids, see Lovins and Lovins, 1982, chap. 10.

3. ARCO Solar has completed a 1-MW photovoltaic plant for Southern California Edison. This plant has been financed privately, and costs have not been disclosed (ELP, June 1982, pp. 58–59).

4. The Electric Power Research Institute is the research arm of the nation's public and private electric utilities. The northern synthetic utility follows the EPRI Summer C scenario; and the southern synthetic utility follows the Summer E scenario (EPRI, 1977).

7
Conclusions

The last decade has been widely perceived as a watershed, an end to unprecedented economic expansion based upon cheap energy. Pessimists envision a human prospect of economic stagnation as the earth's immense storehouse of cheaply extractable fossil energy nears depletion. Even optimists foresee the price of energy rising relative to other commodities and a wrenching and prolonged readjustment of the entire capital stock—buildings, assembly lines, and transportation equipment—to the new realities.

In addition, there has been a heightened awareness of the risks of environmental damage and injury to human health posed by depletable energy systems. These risks are immanent in the entire fuel cycle. Our most abundant fuels appear to pose the greatest immediate environmental risks, particularly in the combustion of coal and in the mining and disposal of radioactive elements. Also, the security of Western democracies is jeopardized by the concentration of the world's petroleum resources in a few unstable Middle Eastern countries.

In the past, the Malthusian specter of resource exhaustion has been banished by technology. Can solar energy provide a technological fix to today's energy problems? The fuel is free. The technologies for extracting the energy directly from insolation, or from its indirect manifestations, like the wind or plants, appear to be benign environmentally. The energy is homegrown and cannot be extinguished by a hostile potentate. So why not a solar fix?

The litany of horrors from conventional energy sources does not point to any particular solar technology as the obvious solution by default. For the next generation, several other options are available: living with the hazards of coal and nuclear energy; spending money to reduce these hazards; encouraging tertiary recovery of domestic oil; or redoubling development efforts in areas like shale-oil production. Solar energy will have to compete with these options as well as alternative potential sources of inexhaustible energy like fusion.

All solar-energy technologies have problems of their own. In decreasing order of severity, they are: the capital cost of energy-conversion systems, asynchrony between supply and demand for energy, limited abundance, system reliability, and potential environmental hazard.

The biggest obstacle to the adoption of solar technologies is their capital costs. Currently technologies for exploiting solar energy are not cheap. Based on their peak capacity, photovoltaics cost about $15,000/kw in 1980, solar-thermal power towers about $6,500/kw, windmills about $2,000 to $4,500/kw, and wood-burning power plants about $2,000/kw. In contrast, coal-fired power plants cost about $800/kw. A solar future, then, entails a more capital-intensive energy system than at present. More relevant is the cost of solar energy compared to the cost of conventional energy replaced. The cost of solar energy is not cheap compared to the alternatives even though its fuel is "free." Economics limits the amount of the abundant solar energy that can be justifiably exploited.

To a society accustomed to energy on demand, certain forms of renewable energy are problematic. The sun and wind do not perform to human whim. Their cycles differ from the cycles of human demand. So long as other energy technologies provide a source of backup energy and a sink for surplus solar energy, the asynchrony between supply and demand may be little problem to the end-user. At low levels of solar penetration, interchanges with conventional utilities solve the asynchrony problem. A society that wished to base its energy system entirely on wind or on photovoltaics would have to confront the storage problem. This problem is not unique to solar energy, but endemic to the electric utility industry, which cannot easily inventory its product.

Innovations in storage technologies would be beneficial to conventional electric systems in improving the utilization of existing base-load capacity. Innovations in electric storage, moreover, would increase the capacity savings from WECS or photovoltaics, which could serve as peaking units. At the moment, pumped water and compressed air appear to be the most attractive storage options. North American utilities are planning to expand their pumped storage capacity by about 50 percent in the 1980s, a bit faster than total generating capacity (NERC, 1983). For regions where land is flat or scarce, this option is expensive; and alternative thermal, chemical, and electric storage systems are even more expensive.

Limited abundance is characteristic of several manifestations of solar energy. Both water power and wind power are immense in a few favored locations. Relative to global demand, hydroelectric resources and wind resources are small even without taking economics into consideration. Even if all the crop residues, timber wastes, and municipal wastes were recycled without regard to economic considerations, they could provide only a small fraction of current world energy demand. If all the growth on crop land and forests in the United States were devoted to energy production, they could meet only about one-third of current American energy demand. The search for a form of energy that is both renewable and virtually unlimited must focus on technologies that exploit insolation directly, like solar-thermal and photovoltaic systems.

The most promising solar technologies are relatively immature. The reliability of these technologies is a major source of risk to potential adopters. Some unknowns that the prospective buyers of equipment face are durability or longevity, the failure rate, and maintenance cost. As with other complex equipment, longevity and failure rates are most likely related to maintenance practices. Currently available solar systems have the high failure rates normally associated with an infant technology. In the simulations here, we assumed that mature technologies could achieve a durability of twenty years, low failure rates (1 percent for photovoltaics, 5 percent for wind-energy conversion systems), and annual maintenance costs of only 1 to 2 percent of initial capital costs. Assumptions about maintenance and reliability are crucial, for the analysis of solar-thermal systems indicates that abnormally high maintenance expenses can erase any benefits in conventional fuel savings.

Environmental risks are probably the least of the obstacles. Nevertheless, hydroelectric dams have often been opposed by environmentalists for their impact on local ecosystems. Wood-burning societies have been plagued by severe air pollution, deforestation, soil erosion, and flooding. The environmental problems posed by solar-thermal systems, windmills, or photovoltaic arrays appear to be minimal. Yet can a society that was unaware of acid rain ten years ago be confident that it can foresee environmental risks? Might the absorption of insolation by thermal and photovoltaic collectors reduce the earth's capacity to reflect insolation back into space? Might a reduction in albedo lead to serious long-term heating of the earth? Might the disposal of toxic chemicals from discarded solar cells, like gallium arsenide or cadmium sulfide, prove hazardous? Surely we cannot answer these questions today, but we must expect some unhappy surprises in this quarter.

Stated plainly, no form of solar energy provides a technological fix to the energy problems here and now. We are faced only with hard choices among risky depletable and risky renewable-energy options. Rather than seeking a panacea today, we can attempt to identify niches where various forms of solar energy can fit in a multifuel energy system. In the near term, where can we expect solar contributions to the world's energy supply? What technical achievements from research and development are required for solar energy to make major contributions in the long run?

What Is the Value of Solar Energy?

Much of the analysis compared the cost of solar energy with the value of the conventional energy displaced. The analysis revealed the following rules of thumb:

- Don't waste time seeking spurious accuracy. There are too many uncertainties in raw data and in assumptions to go beyond two-digit accuracy.
- Focus on the most sensitive influences on costs and benefits, like fuel escalation and discount rates.
- Don't forget "hidden" costs like maintenance or land. High maintenance costs can cancel the annual benefits of a solar energy system.
- Tax status is everything. How you evaluate solar investments depends upon your qualifications under the tax code.
- Don't rely on current market prices in volatile and politically controlled markets, both domestic and foreign. Conventional energy should be evaluated under a probable scenario of complete deregulation of oil and gas and maintenance of tight environmental regulations on coal and nuclear fuel. Surcharges of about 25 percent for environmental and national-security externalities should be considered in social cost-benefit analysis.

The substantive results are straightforward. For this decade, the most promising solar technologies are low-temperature domestic hot-water and wind-energy conversion systems. While photovoltaics will continue to penetrate remote markets in the 1980s, such as irrigation, mass markets must await significant cost-reducing improvements. Biomass is a dark horse.

Domestic hot-water heating comprises about 4 percent of national energy demand in the United States. Well-functioning systems can deliver energy at a social cost of about $20/MBtu in the arid Southwest, about $24/MBtu in the humid Southeast, and about $35/MBtu in the North. Under a social cost-accounting framework, solar hot-water systems are barely competitive with electricity delivered at $18/MBtu in the sunny Southwest. They cannot compete with oil, which can be delivered at $9/MBtu, or natural gas, even if gas were deregulated completely and delivered at $7.50/MBtu. (The escalation of conventional energy prices will raise the levelized cost of fuel savings by 20 to 40 percent above these current values.) In the eyes of homeowners, federal and state income-tax credits can make these systems competitive with oil heating, but not gas heating, in the sunnier regions. For solar hot-water systems to become more than marginally attractive in most of the United States, their costs would have to fall by about 50 percent. Because the underlying plumbing technologies are relatively mature, there appears to be little prospect for such cost reductions.

Residential/commercial space heating and industrial-process steam comprise 15 percent each of American demand. The cost of meeting these higher temperature demands by solar heating is even higher than the cost of domestic water heating. The costs of medium- and high-temperature systems would have to fall by a factor of four or five to be competitive.

A traditional source of mechanical energy, wind is undergoing a revival as a source of electricity. Medium-size wind-energy conversion systems are cost-effective today in a few favored locations where wind speeds average more than 14 MPH and where utilities are heavily dependent upon oil. Where wind speeds average about 16 MPH, wind machines currently on the market can generate electricity for a levelized social cost of about 5.5 cents/kwh. In the United States, promising sites include Hawaii and coastal areas of the Northeast. Many island and coastal locations in the world fit these specifications as well. With a little battery storage, windmills may play a role in small-scale electrification in portions of developing countries that are remote from electric grids.

The potential for extensive market penetration for wind-energy conversion systems is difficult to assess because of the paucity of adequate long-term wind data. Few mainland cities in the United States have measured wind speeds that average as high as 14 MPH at standard height (30 feet). The seventh-power law suggests that doubling the tower height can raise the average wind speed at the windmill hub from 14 to 16 MPH, which appears to be the break-even wind speed for oil-burning utilities. Furthermore, careful wind prospecting may uncover especially windy sites in normally calm regions.

Prospects for reducing the cost of wind-energy conversion systems are mixed. Today, some systems can achieve two-thirds of the theoretical maximum efficiency at certain speeds, leaving little room for improvement. Economies of mass production and improvements in materials may reduce windmill costs by a factor of two, although installation costs are unlikely to fall much. A twofold cost reduction, however, may render wind power economically viable in more than the half of the United States, where wind speeds exceed 12 MPH.

Converting sunlight directly into electricity through photovoltaics offers the greatest potential for a solar fix. Today photovoltaic electricity costs about $1/kwh, which is exorbitant for all but a few remote applications. Photovoltaic technology, however, is rushing ahead on the coattails of the semiconductor industry. If scaled up to mass production, technologies on the drawing board may offer electricity at a cost of about 10 cents/kwh in areas of moderate sunshine and for considerably less in very sunny areas by 1990. These prospective costs are not cheap and offer no illusion of a return to the golden age, but at least they provide a ceiling for electricity prices. Compared to wind-generated electricity, photovoltaic electricity at these prices would appear in almost unlimited abundance in comparison to energy demands.

Because evaporation is proportional to insolation, photovoltaics appear promising for irrigation. Since they have no moving parts and suffer fewer fatal component failures, photovoltaics have a maintenance advantage over windmills in this niche. This promise may be realized as early as the late 1980s in the Southwest. Photovoltaics may also be profitable in residential applications by 1990 in sites with high electricity costs and favorable insolation. Electric utilities that currently burn considerable gas and oil may find photovoltaics attractive in central power stations by the end of the decade as well. Like windmills, photovoltaics offer a promise of village electrification. All of these projections, of course, are contingent upon the realization of the considerable potential for progress in photovoltaic technology.

Electricity provides about 20 percent of American energy demand. Is this figure the upper limit for windmill and photovoltaic penetration? Not at all. The long-term trend toward electrification is likely to continue. Electricity can technically meet the entire gamut of nonmobile energy demands and of some transportation demands through electrified railways. For thermal energy, the choice between electricity and on-site fuel combustion boils down to cost. Centralized utilities can handle dirtier and more hazardous but cheaper fuels more easily than most small end-users. This means that the prices of increasingly scarce oil and gas will rise faster than coal-generated electricity. Furthermore, new high-technology industrial processes, such as high-intensity heating or laser welding, can only be undertaken electrically or electronically (Schmidt, 1983).

The Effect of Tax Incentives

Private analysts of solar-energy investors face market prices, income taxes, tax credits, and depreciation allowances that do not enter into the cost-benefit calculus of the public-policy analyst. The key question is whether the tax structure guides private investors to accept projects that are profitable from the perspective of social cost-benefit analysis and to reject those that are not.

The American tax structure places investor-owned utilities in the least favorable position to profit from solar investments, and it places homeowners in the most favorable position. The tax treatment of utilities is so unfavorable that they would decisively reject solar investments that had social cost-benefit ratios as high as 1.3 to 1. The tax treatment of third-party investors is consistent with a social cost-benefit analysis that levies a 25 percent surcharge for externalities on the market prices of fuels. Homeowners are treated as if these externalities imposed a surcharge of about 40 percent. The tax structure is clearly biased toward decentralized ownership, if not siting, of renewable-energy facilities.

A major policy question is whether such a "small-is-beautiful" bias is justified. Treating the question as an equity issue provides no easy answer. "Homeowners" may appear to be a broader-based constituency than "utilities," who are blamed for escalating fuel rates, overexpansion of nuclear facilities, and acid rain. At the margin, it is not necessarily "fair" that generous residential tax credits are in fact utilized by high-income taxpayers or by third-party investors in high tax brackets, while such credits are denied utilities that serve the entire population.

The bias against utility ownership has nothing to do with the issue of costs of centralization versus decentralization in siting. If they were superior to centralized facilities, distributed photovoltaics could be owned by utilities, third parties, or homeowners. In location or real costs, there is little distinction between a wind farm owned by a utility and one owned by a third-party investor.

The bias against utilities may reflect the same basic Jeffersonian ideology that elicits nostalgia for family farmers and support for small business. There is a profound logic to the Jeffersonian bias: Smallness provides open access to experimentation in energy systems. Revealing biases of their own, utilities have resisted the adoption of energy technologies that are dispersed and those over which they have limited control. The Electric Power Research Institute (EPRIJ, July/August, 1983, p. 2) maintains that for photovoltaics, the "real thing" is bulk electric-power generation. Even here, their prototype wind and solar plants and the establishment of rate schedules for cogenerators were encouraged by external pressures, not by internal demands.

These latter arguments are simply a warning that innovation may not come from leaders in the field, just as the automobile was not invented by the railroad industry. The argument, however, does not imply that the tax structure should discriminate against utilities.

A second major policy question is whether environmental and national-security externalities lie within the 25 to 40 percent range, above and beyond the replacement cost of fuels. Current tax policy implies that they do. If not, then tax credits should be altered in the appropriate direction.

Prospects for Technological Innovation

To some advocates, a solar future epitomizes a new lifestyle, based upon small-scale, community-controlled, "appropriate" technology. Solar energy has been espoused as a soft path toward the future. A consideration of the obstacles on the road to a solar future dispels this notion. Indeed, some solar-energy systems embody low technologies and are amenable to dispersed adoption. These include residential passive design and simple domestic hot-water systems. Such technologies, however, will make relatively minor contributions to the world energy balance. Significant contributions can be made only by mass-produced energy devices that are based on high technology. Only through the realization of economies of scale and through the exploitation of new materials can the costs of windmills and photovoltaics be brought down.

The prospects for cost reductions in photovoltaics are great, but there are many unresolved engineering questions about how best to exploit the photovoltaic effect. A wide range of semiconductor materials, methods of cell manufacture, and system design concepts remain under active consideration. The only agreement is that significant cost reductions can be achieved at annual production levels in single factories that far exceed current world demand.

Technological innovation does not just happen because it is feasible. It happens because manufacturers perceive a profit opportunity in devoting resources to developing a concept into a product. In dealing with photovoltaic technologies, potential manufacturers face several risks. First, there is the possibility of fundamental breakthroughs that could render any manufacturing process obsolete. This militates against the premature commitment of funds to manufacturing. But by holding back, a manufacturer foregoes the opportunity of building up a reputation and brand-name identification. Second, because the achievement of significant cost reductions requires scales of output far in excess of current demand, manufacturers face risks of overexpansion and excess capacity. Third, if potential consumers anticipate continuous technological breakthrough, they have an incentive to wait rather than to purchase now.

While large, these risks are not different in kind from those experienced by other industries in their innovative stage. An energy industry that can sink $1 billion into a nuclear power plant or $8 billion into the Trans-Alaskan Pipeline clearly has the means to invest a fraction of that in a large-scale photovoltaic factory. With great financial staying power, photovoltaic subsidiaries of oil or electronic giants can speed up the realization of market demand by forward pricing; that is, pricing on the basis of anticipated costs.

Integrating Solar Energy with Utilities

While the achievement of cost reductions in high-technology solar-energy systems may require centralized production, their use does not have to be centralized. Some advocates have favored solar energy because of its perceived incompatibility with centralization, and many utilities have feared solar energy for the same reason. Mass-produced solar-energy devices are amenable to dispersed, stand-alone use, but they may diffuse more rapidly in urbanized societies if they can use the central grid as a source of backup energy and as a sink for surplus energy.

Because of the apparently erratic behavior of wind patterns and sunlight, these sources of energy have been viewed by some utilities as destabilizing. Since their charters require utilities to offer reliable service, any source of unpredictability is suspect. Also, utilities have looked askance at solar energy as a "mere fuel saver," offering little reduction in capacity requirements. Are these perceptions accurate?

The issue is measuring "avoided cost" from solar energy. Fuel savings were estimated by a microcomputer-based production-costing model and capacity savings by a loss-of-load probability model. Both models were applied to wind farms and photovoltaics at several penetrations and in five sites, with widely varying climatic conditions and utility characteristics. In addition, the models were tested for two synthetic coal-burning utilities.

Fortunately, most of the fears of utilities are unfounded. The integration of wind-energy conversion systems and photovoltaics with electric utilities poses relatively few problems. Loads on conventional generators are unlikely to become significantly more volatile and hence unpredictable. In some sites, solar or wind penetration reduces the volatility of daily loads.

Most important, both wind and photovoltaic systems save a disproportion of expensive peaking fuels—oil and gas. The higher the penetration of renewables, however, the lower was the share of expensive fuels saved. The "mere fuel savings" from windmills and photovoltaics may more than offset their costs by 1990 in some sites with only average wind speeds or insolation. In sites with superior wind speeds or insolation, profitability comes sooner.

As utilities suspect, solar- and wind-energy systems probably will not reduce the peaks of their current load profiles very much. But in all cases there were substantial reductions in loss-of-load probabilities. Depending upon the capacity-planning rule used, as little as 2 to 4 kilowatts of renewable energy capacity could displace 1 kilowatt of conventional capacity in some utilities. For other utilities there was virtually no capacity savings. The practicality of displacing oil- or gas-fired capacity was questioned, since most of these units are economically obsolete today at current fuel costs. Utilities have no plans to build such units today, and they could find few buyers to share in existing oil and gas plants.

Solar Energy as Social Insurance

The technological characteristics of solar-energy systems pose great uncertainties. Within the decade, most of the uncertainties about durability, reliability, and maintenance will have been resolved. If expectations about the performance and cost can be specified fairly accurately, an investor can view a wind or photovoltaic system as a form of insurance. From an investor's viewpoint, most of the uncertainties about solar-energy systems depend upon the scenarios for conventional energy saved. If fuel escalates faster than expected, an investment in solar energy yields a higher than expected return. The contrary holds when fuel escalates slower than expected. The net effect of solar-energy technologies is to reduce the variance in potential fuel-bill scenarios.

Solar technologies clearly have some value as social insurance. This means that they may be justifiable even if their costs are somewhat more than expected fuel savings. Electric utilities undervalue this insurance characteristic because of the fuel-adjustment clause, which completely insulates utilities from the vagaries of fuel prices and passes the costs on to ratepayers. This divergence in perceptions of utilities and ratepayers suggests that public utility commissions have a legitimate interest in encouraging utilities to exploit this form of social insurance.

Solar Energy in a Transition to a Postpetroleum Economy

The rising costs of oil and gas have been widely viewed as a boon to the economic prospects for solar energy. Indeed, were other things equal, this would clearly be the case. Other conditions do not remain equal, however. In response to rising oil and gas prices, electric utilities are planning to expand their coal-burning capacity. Many utilities in the Midwest are likely to burn very little oil and gas by 1990. By the year 2000, most utilities will be in a similar situation. In this light, the conclusion that wind machines or photovoltaics are likely to save fuel but little capacity becomes more critical. The transition to a coal-fired future raises the possibility that the avoided fuel costs from wind-energy conversion systems and photovoltaics will decline.

Currently, coal delivered to the utility costs about $1 to $2/MBtu. Even if this price rose to $3/MBtu by the year 2000, the value of fuel savings will be relatively low. Translated in other terms, the value of fuel savings may be only 2 to 3 cents/kwh, compared to the 5-cent levelized cost of wind-energy systems in very windy sites. The cost of coal-fired capacity is considerably greater than the cost of oil- and gas-fired capacity, but loss-of-load simulations indicate that these renewable-energy systems may save little coal capacity.

This pessimistic scenario may be altered by two trends. First, utilities are likely to become more sophisticated in load management. They may use techniques like time-of-day pricing, interruptible rates, or remote control of end-user equipment. Or they might be able to inventory energy supplies by some advanced storage technique. If either were the case, then utilities could shift loads to meet the availability of energy from renewable systems. Since insolation and wind speed follow systematic cycles, managing loads around these energy sources is not beyond the capability of utility dispatchers. By shifting loads to the peaks of energy generation from wind or photovoltaics, utilities could then save capacity. In a coal-fired future, these capacity savings could assure the viability of these renewable-energy systems.

Second, the environmental consequences of coal combustion, particularly acid precipitation, may result in stricter air emissions standards. Power plants can meet these standards in two ways: (1) by capturing a larger percent of pollutants after combustion, or (2) by using cleaner feedstocks. The Clean Air Act emphasizes the first path, with its "80 percent reduction" standard, which ignores the original cleanliness of the coal or oil feedstock. This policy has a mixed effect on avoided costs.

Reducing allowable emissions of sulfur dioxide from 1.2 lbs./MBtu to, say, 0.5 lbs./MBtu by installing a superior scrubber translates into a reduction in the environmental externality surcharge but an increase in the capacity credit. Since fuel savings are quantitatively more important by far than capacity savings, the net effect of a stricter Clean Air Act is to reduce the benefits of solar energy. This paradox can be resolved by replacing the percentage-reduction standard with an emission standard.

This strategy would result in utilities bidding up the price of coal, which is either naturally clean or which at some capital cost is precleaned. If this alternative strategy were pursued, the avoided costs of solar-electric systems become much more valuable. The political obstacles to greater utilization of clean Western coal and less dirty Eastern coal are substantial (Landsberg, 1979, chap. 9), so the exploitation of the environmental advantages of solar energy is not a foregone conclusion.

A parallel argument can be made for the national-security benefits of solar energy. The more successful the industrial democracies are in reducing their vulnerability to oil embargoes, the lower is the value of the national-security premium. The diversification of sources of oil, the shift to coal, and the filling of strategic petroleum reserves effectively reduce the value of the energy savings from solar energy.

Policy Recommendations

The research for this book began under the Carter Administration, which proclaimed the goal of a "20 percent solar America" by the year 2000, and was completed under the Reagan Administration, which is dismantling federal solar-research programs. Does this change in emphasis bode ill for the solar prospect?

Not at all. The groundwork has been laid. The decontrol of oil prices and the partial decontrol of gas prices is bringing the price of conventional fuels closer to their marginal social cost. The generous federal income-tax credits remain intact. The residential tax credits are threatened, but provisions of the Economic Recovery Act of 1981 make alternative-energy investments far more attractive to business. Market developments are likely to induce commercialization of those solar technologies that seem most promising in the near future. Indeed, the photovoltaic industry in the United States, Europe, and Japan is building pilot plants now. Wind farming is an emerging business, particularly in California and Hawaii. In sum, the accumulation of policy initiatives and market changes over the last decade seems to be in the right direction.

What other policy initiatives need to be taken?

1. Energy prices should be brought even closer to their marginal social cost. This policy has three facets:

- The well-head price of natural gas should be fully deregulated. Whether or not a windfall profits tax is levied on producers of "old" gas is irrelevant to the stimulus provided to conservation and the adoption of solar-energy systems.
- A national-security surcharge should be imposed on oil and gas

imported from overseas. This surcharge may take the form of a tariff or a compulsory deposit in a stockpile. Suppliers from secure Western Hemisphere sources should be exempt from this requirement. Oil and gas prices in the United States should be allowed to rise to meet the new imported price.
- Charges should be levied on emissions of oxides of sulfur and nitrogen, which have serious health and environmental consequences. A research program on computing the cost of carbon dioxide emissions should be initiated with an eye to levying a "carbon tax" or stricter regulation.

2. The federal government should maintain current incentives for research and development expenditures by private industry. Because the scientific principles of solar thermal and wind-energy conversion are well known, no public funding for research and development of these systems is called for. Current tax incentives appear to be sufficient to encourage the funding of basic research on photovoltaics and biomass by private industry. (The electric utility industry has sufficient funding through EPRI to support research on integrating renewable-energy systems with the grid.) As these technologies approach the commercialization stage, it is appropriate that the lion's share of development expenditures be funded by the private sector.

3. Both the public and private sectors have a role in enhancing the credibility of solar technologies. Credibility is based on perceptions of cost and performance. While the citizenry may favor solar solutions abstractly, they may be reluctant to make a costly investment in a solar system because of doubts about performance, including durability, reliability, and maintenance costs. It is critical that these doubts be laid to rest by a concerted effort at field testing. If the solar industry expects to penetrate the mass market by the late 1980s, such testing should begin now. In its own interest, the solar industry should submit its products to be tested under impartial auspices. The most credible would be independent bureaus experienced in testing electrical equipment, national laboratories, or state solar-energy demonstration centers.

4. Solar tax incentives should be reassessed when the current law expires in 1985. The currently generous tax credits for alternative-energy investments may become increasingly unnecessary and increasingly counterproductive in the late 1980s. While they are useful in helping an infant industry achieve economies of scale, they can also degenerate into protection for inefficient, special interests. The credits should be phased out gradually in the 1985–90 period. In the interim, utilities should become eligible for the same generous tax credits and accelerated depreciation allowances for alternative-energy systems as other businesses.

5. Current laws ensuring the rights of solar cogenerators should be enforced. In particular, utilities should be required to pay "avoided costs" to cogenerators. While not as easily calculated as in the economics textbooks, the avoided costs are likely to differ substantially from the average rates currently charged by most utilities. The value of electricity sold back may be especially critical to the profitability of decentralized photovoltaic and wind systems. The right to sell electricity to utilities at avoided cost will foster a wide range of experiments in different parts of the country with a variety of solar-energy systems. As soon as utilities acquire sufficient experience with interactive solar-photovoltaic and wind-energy conversion systems, these laws should be allowed to wither away. If utilities find solar cogeneration in their interest, the laws become unnecessary. If solar cogeneration proves to be uneconomical for either the utility or the cogenerator, the laws are undesirable.

6. The value of solar energy as social insurance should be recognized. The automatic fuel-adjustment clause should be modified so that utilities have an incentive to consider this value in their investment decisions.

Solar photovoltaics may not offer a guaranteed fix to our energy problems, but it does offer a very promising option. The solar option is fraught with uncertainty, and there may be unpleasant as well as pleasant surprises as this option unfolds. The private sector perceives the promise and is investing increasing sums in product development. The public sector is supporting these private developments by generous tax incentives to consumers. Consumers are beginning to risk substantial sums on these expensive systems. The solar option is risky, but it is a risk bearing hope.

Appendix: Parameters for Cost-Benefit Analysis

By now, a rather large body of unpublished literature evaluating the economics of solar-energy systems has accumulated (such as Doane et al., 1976; Perino, 1979). The conceptual diversity of this literature is disconcerting. Analyses purporting to answer the same questions will more often than not utilize different vocabularies and different arithmetic manipulations. While most of these analyses are correct, the superficial variations mask the basic similarities in approach. For example, what some analyses call a discount rate others term an interest rate, a perfectly substitutable synonym. Some analyses deal with annualized costs, while others deal with capitalized benefits, a distinction in arithmetic manipulation that makes no difference in the ranking of alternative solar systems. Some concepts, such as levelized costs, originate in the utility industry, while others, such as discounted cash flow, originate in finance, and still others, like discounted social benefit, originate in economics. Most perplexing is the widespread use of highly contorted figures of merit and mathematical manipulations when simpler ratios would suffice.

The key issues in evaluating any investment are establishing the investment criterion, specifying financing terms, correcting for inflation, and choosing a discount rate. An issue specific to solar-energy investments is projecting the escalation of conventional fuel prices.

Investment Criteria

Compounding and Discounting

Compounding and discounting are fundamental concepts in the analysis of investments. These concepts are widely misused, yet they are basically simple to understand. We can first show how discount rates are used to compare economic values in different time periods and then consider where these rates come from. In the days of stable prices, the interest rate represented the increase in purchasing power that could be obtained by depositing one dollar in a bank. Leaving B dollars in a bank for t years resulted in a compounding of interest, such that

$$B_t = B_0(1 + i)^t$$

where B_t is the future value of deposits; where B_0 is the initial deposit; and where i is the interest or discount rate.

This compounding formula can be used to solve the following problem. If you borrow $10 at 5 percent interest, with interest and principal payable at the end of twenty years, how much will you have to repay? The answer can be obtained by punching a few keys on a desk-top computer. For annual interest payments, the compounding factor for 5 percent interest for twenty years is 2.6533. In the example, the future value of $10 is $26.53 (10 × 2.6533).

Discounting, the opposite of compounding, indicates how much a given sum in the future is worth today at a particular interest rate. The reciprocal of the compounding factor, called the present worth factor, indicates how much one dollar twenty years in the future is worth today at 5 percent annual rate. The reciprocal of the compounding factor of the example above yields a present worth factor of .3769 at a 5 percent discount rate and a twenty-year term. The present worth of $26.53 is found either by dividing this sum by the compounding factor or multiplying it times the present worth factor. In either case, the answer is the same: $10.

$$B_0 = B_t/(1 + i)^t$$

While compounding brings some present value into the future, discounting brings some future value into the present. By the vehicle of an interest or discount rate, economic values at different times are made commensurable. In the example above, discounting and compounding occurred on an annual basis. Financial intermediaries can compound or discount on a quarterly, monthly, or daily basis as well.

Capitalization involves discounting an entire stream of future benefits or costs to the present. Arithmetically, the present worth (PWB) of the stream of benefits received for T years can be expressed as:

$$PWB = B_1/(1+i)^1 + B_2/(1+i)^2 + \ldots + B_T/(1+i)^T \text{ or}$$

$$PWB = \Sigma B_t/(1+i)^t$$

When the annual benefits B are equal in each year, this expression reduces to a simple formula:

$$PWB = B_0(1/i - 1/i(1+i)^T)$$

When benefits escalate at a constant rate r, the calculation is a little more complex. In practice, one can approximate the present worth by subtracting r from i and then simply discounting as if benefits were level. For example, assume that a 5-kw residential system saves $1,000 in electricity bills each year. Annual maintenance costs are $100. With a discount rate of 5 percent, the present worth of savings is $12,462 and the present worth of maintenance $1,246. In capitalizing, it may be more convenient to subtract annual maintenance costs from gross annual benefits before capitalizing rather than capitalizing each separately. Program PW SAVINGS provides a simple algorithm for calculating the energy savings, net of maintenance.

The cost of solar-energy systems is expected to decline over time. For example, photovoltaic array costs may fall 26 percent per year, while balance of system costs may fall 20 percent per year for the next decade. In other words, the cost of a system of design capacity h has a cost that depends upon the year of investment j:

$$COST_{hj} = ARRAY_h\, e^{-.26j} + BOS_h\, e^{-.2j}$$

When taxes are taken into account, the costs to utilities, third-party investors, and homeowners deviate somewhat from COST. If the investor pays part of the COST from equity (retained earnings or stock issues) and part by borrowing, the present worth of capital costs to the utility equals:

$$COST(U) = EQUITY + \Sigma\,((1-c)\,I_t + (A_t - cD_t)/(1+i')^t - TC\,(COST)$$

where TC is the investment tax credit; where c is the marginal income-tax-rate percent; where A is amortization on the mortgaged half of the investment; where D is depreciation; and where i' is the cost of capital. The tax credit is received for the entire cost, regardless of financing.

On the assumption of 100 percent financing, the cost of solar systems to homeowners can be expressed as:

$$COST(H) = \Sigma\,((1-p)\,I_t + A_t)/(1+i(1-p))^t - TC\,(COST)$$

where p and TC are personal income-tax rates and tax credits, respectively, and i is the market interest rate.

The net present worth of a system purchased in year j can be expressed by the formula:

$$NPW = PWB - COST$$

Program NPW RESIDENTIAL calculates the present worth of residential solar systems and the optimal time of investment from several investor viewpoints. The program accepts alternative discount and fuel-escalation scenarios.

Levelization

Levelized systems costs are another indicator of the profitability of solar investments. Levelizing or annualizing is just the opposite of capitalizing. The levelized cost is a hypothetical level stream of annual costs LAC with the same discounted present worth as the capital cost. The levelized cost of a solar-energy system meets the condition:

$$COST = \Sigma LAC/(1 + i)^t$$

This expression has the solution:

$$LAC = COST\, (i/(1 - (1+i)^{-T}))$$

The factor that is multiplied by COST is called the capital-recovery factor and can be found in standard financial tables. In the example of a $10,000 system, the capital-recovery factor is .08026, and the annual payment necessary to repay principal and interest at 5 percent is $802.60.

Other annual costs, such as insurance and property taxes, are proportional to the cost of the capital equipment. The total levelized cost of capital—the sum of capital recovery, insurance, and property taxes—is called the fixed-charge rate. In the hypothetical example, if insurance and taxes amount to 1 percent of capital costs, then the fixed-charge rate is .0902 (.01 + .0802).

Levelized costs can be expressed in terms of kilowatt-hours. If the hypothetical system saves 20,000 kwh per year, the levelized cost of solar electricity is 4.5 cents/kwhs (902/20,000). Program LEVELIZATION performs these calculations for renewable energy systems.

Overcoming Financing Obstacles: Creative Financing

While any stream of benefits or costs has a unique capitalized value at a given interest rate, a given discounted present worth is consistent with many streams of benefits and costs. This fact of arithmetic enables financial intermediaries to design innovative schemes for financing solar systems.

Suppose we identify a profitable solar investment, like the optimal system identified for New York—25m² thermal/75m² photovoltaics. In the first year of operation, will the cost of this system be less than the value of conventional energy saved? Behind this seemingly straightforward question is the concern that unless homeowners obtain visible energy savings in the first year they will not purchase the system.

The question assumes that the solar system will be financed by a level mortgage, which until recently was the mainstay of housing finance. Under a system of level mortgage financing, a regular payment, MORT, is found that will repay the principal plus interest over a given term. As indicated in the discussion of levelization, the financial institution finds the value of $MORT_t$ that solves the equation:

$$COST = \Sigma\ MORT_t/(1 + i)^t$$

where $MORT_t$ is constant for all t.

If the system was being leased and financed by a utility, the utility would find a lease payment LEASE that solved the equation:

$$COST(1 - TC) = \Sigma\ (1 - c)\ LEASE_t/(1 + i)^t - \Sigma\ cD_t/(1 + i)^t + \Sigma\ (1 - c)\ MAINT_t/(1 + i)^t$$

where TC is the tax credit; where c is the marginal corporate tax rate; where D_t is the depreciation allowance; and where $MAINT_t$ is the maintenance charge. Instead of a lease payment, the utility might charge a fee, $RATE_t$, for each kilowatt-hour of electricity generated. In this case the utility would solve the simple equation:

$$RATE_t = LEASE_t/KWH$$

where KWH is the annual expected electric output of the system.

So far the arithmetic is simple, but there is no guarantee that in the first year $MORT_t$, $LEASE_t$, or $RATE_t$ will be less than the conventional electricity savings. Fortunately, we can exploit the arithmetic fact that there are an infinite number of streams of mortgage or lease payments whose value equals COST. So long as the solar investment is profitable, we can escalate mortgage or lease payments at the same rate as conventional energy costs while keeping these payments below conventional energy savings every single year of the investment.

For example, in 1990 the optimal system in New York City costs $13,100, saves 8732 kwh, and permits sellback of 4365 kwh each year. Weighting sellback by .9, the electricity output is 13,097 kwh per year. Putting the proper parameters in the LEASE equation, one solves for an annual lease payment of $1,200. From the RATE equation, one obtains a level electric charge of 9.2 cents/kwh. This compares to the average value of electricity saved and sold in 1990 of only 8.9. In other words, the first-year savings would be less than the cost of a level lease payment.

Instead, the utility might charge a price that escalated with conventional rates r. Then the problem reduces to one of finding P_0, the initial price in 1986, which solves the equation:

$$\Sigma P_0 e^{rt} \text{KWH}/(1+i)^t = \Sigma \text{LEASE}_t/(1+i)^t$$

If the anticipated escalation rate is 2 percent, the initial price P_0 must be 8.1 cents/kwh. If anticipated fuel escalation is 4 percent, then the initial price only need be 7.2 cents/kwh. These are both below the first-year costs of conventional electricity. With creative financing, the utility recaptures its investment, and the homeowner saves not only over the life cycle of the solar investment, but each and every year. Because homeowners are not taxed on energy savings, the lease payments could be even lower if a utility or a financial intermediary lent to the homeowner directly.

How to Treat Inflation

Inflation is the persistent increase in the general price level, as distinguished from a persistent increase in the relative price of a single commodity, like petroleum.

Some analysts of solar-energy systems utilize current or nominal fuel prices in calculating the stream of benefits (Jones, 1982). These prices reflect their anticipation of both the increase in relative energy prices and general inflation. This is an undesirable practice for two reasons. First, economists have neither theories of predicting inflation over a long period nor techniques for doing so. Inflation results from monetary policy, which can change drastically at any moment. Economists therefore generally restrict themselves to short-range projections over the several quarters it takes for current monetary policy to work its effects. Second, if analysts could correctly project future prices, interest rates would adjust themselves accordingly, with no change in capitalized benefits.

This second point merely reflects the rational adaptation to expectations (Roll, 1972; Sargent, 1973; Keran, 1976). Institutions that finance these investments must promise savers a real interest rate i that reflects the expected real increase in purchasing power of savings. Suppose that prices are expected to increase at the rate q over the lifetime of the solar investment. Rational savers will insist upon a nominal interest rate that will protect their purchasing power as well as providing the real rate of interest i. The nominal interest rate must equal q + i. If both benefits and the discount rate are increased by the same q, the net present worth is unchanged. This point can be demonstrated arithmetically using continuous compounding.

$$\text{NPW} = B_0 \Sigma\ e^{qt}/e^{(q + i)t} = B_0 \Sigma\ e^{qt}\ e^{-qt}\ e^{-it} = B_0 \Sigma\ e^{-it}$$

If inflation is correctly anticipated, its inclusion in the calculation adds nothing to the analysis. Because inflation is never correctly anticipated by everyone, inflation adds an element of uncertainty to any investment. Because it is so hard to forecast inflation over the long run, financial intermediaries are beginning to adopt variable mortgage rates, whose nominal value varies with recent changes in the price level in order to maintain the real rate of interest. The widespread use of variable mortgage rates to finance solar investments would render the use of constant nominal interest rates unrealistic.

What Is the Proper Discount Rate?

What is the proper rate for discounting benefits or annualizing costs? How does this rate relate to the bewildering array of interest rates observed in the market? Is a common discount rate used whether the solar-energy system is owned and financed by the utility, owned by the homeowner but financed by a bank or government agency, or owned and financed by the homeowner?

From the private investor's viewpoint, the answer to these questions is simple: Select the interest rate available to you on the market. From the public-policy analyst's viewpoint, no such simple answer is available. Indeed, the selection of the proper discount rate is one of the most difficult problems in social cost-benefit analysis (Lind, 1983a).

Interest Rates in Theory

Discount rates are prices that reflect a person's time preferences. By this is meant their valuation of consumption in the present versus consumption in the future. A person's time preference is indicated by his answer to the question, "How much additional purchasing power next year would induce you to save an additional dollar today?" If the sufficient inducement is 5 cents, then at that level of savings the marginal rate of time preference is 5 percent. If the person was already saving a high proportion of his income, the inducement might have to be larger, say 10 cents. In general, the marginal rate of time preference increases with the amount of present consumption being sacrificed. In other words, the savings-supply curve slopes upward.

Like all prices, market interest rates are determined by supply and demand. While the supply curve reflects the time preference of the savers, the demand reflects the potential return on investments in the economy. The market interest rate and the amount of savings are determined by the intersection of supply and demand curves.

The concept of the "market interest rate" is an abstraction. In fact, there is a plethora of interest rates, and they are related to each other in a systematic manner. Three important factors that differentiate interest rates are liquidity, risk, and taxes.

The liquidity of an asset is the facility with which it can be sold at any moment at the market price. Investors tend to demand lower returns on the more liquid assets. The riskiness of an investment is defined as the variability in returns from one period to the next. In general, riskier investments are discounted at a higher rate. On the one hand, solar-energy investments are fairly illiquid because they are difficult to separate from structures; thus they should be discounted at a premium. On the other hand, solar investments are a form of insurance and thus should be discounted at a relatively lower rate. How these two factors balance has not been determined.

To illustrate the effect of taxes on discount rates, consider the market for relatively safe blue-chip stocks. Suppose that a corporation can earn 10 percent, before taxes, on the last investment it undertakes. If the corporate tax rate is 40 percent, the corporation can return only 6 percent to stockholders. The market return on the stock is thus 6 percent, while the gross return of 10 percent, which reflects the opportunity cost of the investment, is not reflected in the market interest rates at all. In other words, the social return foregone by diverting equity capital into a solar-energy investment is 10 percent. Four points of this return are foregone by the Treasury and six points by the stockholders.

Stock dividends and interest (with the exception of interest on municipal bonds) are subject to the personal income tax. The net aftertax return on savings is reduced in proportion to the personal income-tax rate. For example, if corporate stock yields dividends at the rate of 6 percent and the individual income-tax rate is 33 percent, the net return is 4 percent.

The market determines both the level of investment and interest rate simultaneously. In the hypothetical example, the market rate of return on stock is 6 percent. The social opportunity cost of investments is 10 percent, and the net return on savings, an indicator of consumers' time preferences, is 4 percent. Neither of these latter interest rates is represented in the market.

Let's consider a corporation that owns and maintains solar-energy systems. If it were like the average utility, it would be capitalized with debt and equity in equal proportions. What are the social costs of capital, the corporate costs of capital, and the consumption rate of interest implied by corporate investment in solar systems? The social cost of capital is the weighted average of the pretax rates of return on debt and equity, here 6 and 10 percent, respectively. At an average debt-equity ratio of 1:1, the social cost of capital invested in solar systems is 8 (.5 × 10 + .5 × 6) percent.

The corporation must pay its bondholders and stockholders 6 percent each. Because interest payments are tax deductible, the posttax cost of debt is only 3.6 (.6 × 6) percent. The corporate cost of capital invested in solar systems is 4.8 (.5 × 6 + .5 × 3.6) percent.

Consumers who invested in a solar system would be satisfied with a rate of return equal to the posttax return on their investment portfolio. If this portfolio were invested equally in debt and equity, this return would be 4.0 (6 × [1 − .33]) percent.

In this manner, taxes create a disequilibrium in capital markets. Of the 10 percent produced from each dollar invested, the government receives 4 percent in corporate income taxes. Of the 6 percent returned to corporate stockholders, the government receives another 2 percent in personal income taxes, leaving the individuals 4 percent after taxes. Corporations could produce a return to society of 10 percent for each additional dollar invested, but individuals require only 4 percent to be induced to provide an additional dollar of savings. The wedge between the social cost of capital and the consumption rate of interest indicates that society as a whole would be better off at a higher rate of savings and investment, even though no individual has any incentive to save or invest more. This conclusion is not peculiar to solar investments but is characteristic of the entire investment market.

Lind (1983a, 1983b) indicates that the correct procedure is multiplying the cost of an investment by a factor reflecting the foregone returns and discounting by the consumer's rate of interest. Calculating this multiplier depends upon assumptions about how an investment is financed. The correct range is bracketed by the simpler technique of analyzing investments at both 5 and 10 percent discount rates.

The contribution of solar-energy systems to other public purposes, such as independence from foreign pressures or reducing pollution, does not justify a lower interest rate. External benefits of solar energy should be explicitly added to the annual benefits or, conversely, to the costs of conventional fuels.

This discussion suggests the level of interest rates that government, corporations, and individuals ought to use. But what are the interest rates that each party in fact uses?

Consumption Rate of Interest

The consumption rate of interest is dependent upon the marginal tax rate on personal savings and borrowing. With the progressive income tax, this marginal rate increases with income. Consequently, low-income individuals would be expected to have a higher discount rate than high-income individuals in light of similar market rates of interest. Low-income individuals would thus appear to be more present-oriented and less provident for the future.

Haveman (1969) calculated the aftertax interest rates that consumers in each income category received on their savings and paid on their borrowings under conditions prevailing in 1966. For households earning more than $15,000, then in the top income quartile, the weighted interest rates were about 6 percent. For the entire population, the weighted interest rate was over 7 percent. These nominal rates must be corrected for inflation, which averaged about 1.4 percent in the preceding decade, bringing the consumption rate of interest down below 5 percent for the upper-income group. Since the upper-income group is most likely to provide the market for new housing, and hence for solar-energy systems, a real rate of 5 percent appears to be the relevant consumption rate of interest. These derived interest rates indicate what consumers actually earned (ex-post interest rates), which is not necessarily equal to what they expected to earn (ex-ante interest rates).

Will consumers discount the benefits of energy-saving investments like solar systems at 5 percent? Hausman (1979) presents evidence that consumers utilize much higher rates of discount in practice. Air-conditioner prices vary in proportion to their energy efficiency. On the assumption of an equipment life of nine years, Hausman finds that buying patterns reveal an implicit discount rate of about 15 to 25 percent for all consumers and about 9 to 17 percent for high-income consumers. Reasons for this high discount rate may include consumer ignorance of equipment efficiency ratings, uncertainties of the equipment durability, or uncertainties of trends in future electric costs. Consumers may deal with such uncertainties by implicitly using short time horizons or high discount rates.

Corporate Cost of Capital

Since electric utilities are the businesses most likely to own and lease small-scale solar-energy systems, let's consider their cost of capital. In the period of declining fuel prices that faced the utility industry prior to the 1970s, their securities were relatively riskless, bonds were highly regarded by the credit-rating agencies, and stockholders were assured of steady yields. In the 1970s utilities found that they were unable to obtain rate increases from regulatory commissions in line with increases in fuel and capital costs. Regulatory commissions tended to base the allowable "fair" return on historical or "embedded" interest rates and capital costs, rather than on the nominal values generated by inflation and by the costs of pollution-control equipment. Thus the securities of utilities have become riskier and have had to offer higher returns as a result.

In the period 1955–65, when inflation proceeded at the low rate of 1.4 percent per year, utility stocks yielded about 3.5 percent. Only 60 to 70 percent of utility profits were distributed as dividends (the remainder being retained for reinvestment), so the real rate of return was about 3 or 4 percent (Keran, 1976). In a longer period, 1958–76, the return on book equity was about 12 percent. Since the market value of shares was about twice book value, the real cost of equity to utilities was about 6 percent (Thompson, 1979).

The annual variability in dividends from utility stocks was much greater in the 1960s and 1970s than in the 1950s (Pettway, 1978). As a result of the increasing riskiness of utilities, the average return on utility stock relative to industrial stocks increased over the period 1960–68. The risk premium on utility stocks appears to be about one point above that for industrial enterprises of similar size. The rising cost of equity capital has induced utilities to increase the share of debt and reduce the share of equity in their capital structures. This in turn increases the riskiness of utility stock further (Moody's Public Utilities, 1978).

The cost of debt in the era of stable prices ending in the mid-1960s was about 4.5 percent for Aaa-rated utilities and slightly higher for Aa-rated ones. In the inflationary period following, nominal rates have risen to more than 11 percent for the highest rated utilities. Furthermore, as a result of increasing riskiness, few utilities now receive the highest rating. As a result, the average utility has to offer a real return of 6 percent on its bonds.

The cost of capital to corporations like utilities is computed as the weighted sum of the costs of equity and debt after taxes. On the assumption of a 40 percent corporate income-tax rate, the tax-deductibility of interest on debt results in an aftertax cost of debt of 3.6 percent. The shares of debt and equity in the capital structure are about equal. Thus the weighted cost of capital in real terms is about 5 (.5 × 3.6 + [.5 × 6]) percent. Corey (1983) finds that, in practice, about half of the nation's utilities uses this aftertax rate, but another half uses the higher pretax rate of about 10 percent.

Governmental Cost of Capital

The Office of Management and Budget promulgates an official discount rate to be utilized in evaluating all federal investments. Since 1972, the rate has been 10 percent, but a lower rate, about 7½ percent was allowed for water projects. At that time, long-term U.S. Government bonds yielded approximately 10 percent. With an anticipated inflation of over

5 percent, these rates imply a real federal discount rate of 5 percent for energy projects.

Since Congress has decreed that solar-energy investments fulfill a "public purpose," there should be little question of their eligibility for financing by state or local intermediaries issuing municipal revenue bonds (White, 1979). The highest rated tax-free municipal bonds are yielding about 7 percent, implying a real discount rate of 2 to 3 percent for municipal utilities.

How Fast Will Energy Prices Escalate?

What expectations should a potential purchaser of a solar-energy system have about future energy prices? Is there a "cap" on fuel-price escalation?

Simple Theories of Fuel-Price Escalation

The simplest theory of fuel-price escalation assumes that all fuel deposits have been identified and are of uniform quality (Fisher, 1979; Nordhaus, 1979). Hence, a fixed endowment of resources is to be depleted over time. Fuel deposits are depletable capital, and rents or royalties are the annual returns. These royalties are the value of the right to mine resources and are equal to the difference between the price of the mined resource and the variable costs of mining; that is, reproducible capital and labor.

According to this theory, the value of fuel deposits in the ground, under competitive conditions, is likely to increase at the same pace as the real discount rate. The reason is that in efficient capital markets the rate of return on all assets (adjusted for risk) is similar. If the return on holding fuel deposits fell below the rate of return on other forms of capital, owners of these resources would try to unload them, thereby forcing down the price once and for all. On the other hand, if the return on holding fuel deposits exceeded the rate of return on other forms of capital, owners would bank these resources, thus forcing up the price once and for all. After these adjustments, the return on holding deposits should equal the return on other assets (Solow and Wan, 1976; Heal, 1976).

The cost of delivered fuel depends upon these royalties, as well as the costs of extraction, transportation, and refining. Since transportation itself is an energy-intensive activity, transportation costs should follow the trend in fuel costs. While the royalties escalate at the interest rate, the maximum price the fuel can achieve is that of the next cheapest, or "backstop," technology. Under the current OPEC-dominated price regime, the extraction costs for oil are minimal, around 1 percent of sales price. There is little reason to believe that extraction costs will increase less than the discount rate in the future. Fuel-escalation rates are thus likely to be dominated by increases in resource rents. These price increases enhance the viability of deposits with ever-increasing extraction costs.

According to this simple theory, the price of oil and its close substitutes are likely to escalate at 5 to 10 percent, the real return on capital in the world economy. This simple theory would have fared poorly in explaining fuel-price trends prior to 1973, when prices dropped. Prior to the oil embargo, fuel prices fell because of new discoveries, particularly in the Middle East and North Africa, and improved extraction techniques. The American and European oil companies that developed these oil fields acted as if their time horizon were shorter than implied by the theory.

Under a regime of monopoly, royalties will follow a different time path. Fuel prices will lay slightly below the level of the backstop technology or alternative fuels and will escalate slowly. Prices of the monopolized fuels will equal the price of the backstop technology just as these fuel deposits become exhausted. If one believes that the world petroleum market is governed by economically motivated monopolists, then small fuel-price increases can be expected in the future. Nordhaus (1979) calculates the rise in the monopoly price at about 2 percent a year.

Because the price trajectory of conventional fuels is influenced by the price of the backup technology, research and development decisions in the consuming countries play a major role in setting the height of this trajectory. In addition, this trajectory is influenced by patterns of future exploration and discovery, inventions of cheaper modes of extracting fuels, and the creation of multiple backup technologies (Fisher, 1979). The modeling of energy exploration and discovery remains quite rudimentary.

Complex Models of Fuel-Price Escalation

Several dozen studies have attempted to project energy demand in the context of complex simulation models. Most models explain energy demand as a function of such factors as population, labor force, productivity, and gross national product; the composition of this national product and interindustry linkages; the elasticity of substitution among fuels in production and consumption; exploratory activity; and public policy. Most studies, such as those reviewed in Schurr et al. (1979), assume real annual increases in coal and electricity prices of about 2 percent through the year 2000. Gas prices are assumed to rise at a much faster rate as a consequence of deregulation.

A few simulation studies attempt to project energy prices endogenously, rather than assuming some given rate of growth. Hudson and Jorgenson (1974) developed a world energy model of enormous sophistication for projecting energy use and prices through the year 2000. Following the upsurge of 1973–75, petroleum prices were projected to grow no more than 1 percent in real terms, while coal and gas prices were projected to grow at a rate of 3 percent a year. Electricity prices were projected to mirror inflation, thus evincing no real growth. Despite the refinement of the model, many of the projections have been invalidated by events, for crude petroleum prices were projected to reach merely $11/bbl. in 1985, or $21/bbl. in current dollars by the year 2000.

Several alternative modes of projecting energy prices were utilized in a massive study by the Edison Electric Institute (1976). An econometric demand model of Data Resources, Inc., was supplemented by informed judgments on supply by experts from the Electric Power Research Institute and Edison Electric Institute under three growth scenarios. The high-growth scenario is quite optimistic about the supply elasticities of fuels. Exploration was projected to bring real prices of crude oil down in the 1975–85 period, with prices rising somewhat thereafter. Under this scenario, real prices of oil and natural gas would actually decline over the last twenty-five years of this century. The low-growth scenario is pessimistic about supply elasticities; therefore, decontrol is expected to result in a doubling of real oil and gas prices in the period 1975–85. Thereafter, oil prices are expected to remain steady at $10/bbl. in 1975 prices through the year 2000. The most likely, the moderate-growth scenario, results in the greatest oil price increases as decontrol raises prices at the rate of 3 percent until 1985 and 1½ percent thereafter. Because of improved heat rates and other productivity gains in utilities, average electricity prices were expected to drop by 1 percent per year, to 1.6 or 1.7 cents/kwh in 1975 dollars. All three scenarios have underestimated the speed of decontrol, the ability of OPEC to control prices, and the impact of such political events as the Iranian revolution. Furthermore, coal prices were ignored, a blatant omission in light of the massive conversion of electric utilities to this fuel. Again, these projections are clearly obsolete.

The National Photovoltaic Program (DOE, 1979) projected year 2000 prices of residual oil at $5.14/MBtu and of coal at $3.47/MBtu (in 1980 dollars). At current coal prices of $1.50/MBtu, the projection implies an escalation rate of 3 percent. Real electricity rates are expected to rise from 0 to 3 percent until 1990 and less thereafter.

Subsequent analyses revise expected fuel prices in an upward direction, but insufficiently. Schurr et al. (1979) estimated that the replacement cost of oil would probably plateau at $18/bbl., double the 1978 price. The plateau would be set by the price of nuclear fission and synfuels. Historically, estimates of the costs of nuclear energy and synfuels have been overoptimistic. Unforeseen difficulties have raised the costs of nuclear electricity from being "too cheap to meter" to about the same level of coal-generated electricity, or more (Fenn, 1981). Shale oil was expected to cost only $4 to $6.50/bbl. in the early 1970s; synfuels from coal were expected to cost $8 to $9/bbl. By 1979 these estimates rose to $22/bbl. and $15/bbl., respectively (de Marchi, 1981, p. 7).

A distillation of these studies suggests that the most likely escalation in real world prices of fuels in the next twenty years is about 3 percent, bounded by 2 and 4 percent. Electricity prices are projected to rise at a somewhat slower rate because per-unit capital costs are not expected to continue rising indefinitely, and because utilities are expected to increase fuel-conversion efficiency and to improve capacity utilization.

These economic models neglect political influences on energy supply, such as wars or revolutions in producing countries or decisions on the development of alternative supplies, such as nuclear or solar, in the consuming countries. Because long-run energy supply is heavily infuenced by political events, the validity of these economic models is limited. Clearly, most projections have been wrong. The only alternative is to undertake a sensitivity analysis of solar investments under different price-simulation scenarios.

BASIC Computer Programs

Program PAYBACK

```
 10   TEXT : HOME : PRINT "PROGRAM PAYBACK:": PRINT
 20   PRINT "THIS PROGRAM CALCULATES THE NUMBER OF
      YEARS REQUIRED TO PAY BACK THE INVESTMENT IN A
      SOLAR SYSTEM .": PRINT
 30   INPUT "CAPITAL COST OF SYSTEM= ";CC
 35   INPUT "ANNUAL MAINTENANCE AS PROPORTION OF
      CAPITAL COSTS ";AM:AM = AM * CC
 40   INPUT "FEDERAL TAX CREDIT (AS DECIMAL) ";FC
 50   INPUT "STATE TAX CREDIT (AS DECIMAL) ";SC
 60   INPUT "ANNUAL ENERGY SAVED IN MBTU OR KWH ";ES
 70   INPUT "VALUE OF ENERGY PER MBTU OR PER KWH
      ";VS:EV = ES * VS
 80   INPUT "ENERGY ESCALATION RATE (AS DECIMAL) ";ER
 90   SVE = 0:YR = 0
100   SVE = SVE + ES * VS * (1 + ER) ^ YR - AM
110   YR = YR + 1
120   IF SVE < = CC * (1 - FC - SC) THEN GOTO 100
```

130 PRINT : PRINT "PAYBACK WITH TAX CREDITS IN YEAR ";YR
140 IF SVE > = CC THEN PRINT : PRINT "PAYBACK WITHOUT TAX CREDITS IN YEAR ";YR: GOTO 180
150 SVE = SVE + EV * (1 + ER) ^ YR − AM
160 IF SVE < CC THEN YR = YR + 1: GOTO 150
170 PRINT : PRINT "PAYBACK PERIOD WITHOUT CREDITS IN YEAR";YR + 1
180 INPUT "DO YOU WISH TO TRY ANOTHER SITE (Y/N)? ";X$
190 IF LEFT$ (X$,1) = "Y" THEN GOTO 50
200 REM WRITTEN BY MARTIN KATZMAN IN APPLESOFT BASIC, MAY 1982

Program LEVELIZATION

10 TEXT : HOME : PRINT "PROGRAM LEVELIZATION: (VERSION AUG. 1983)": PRINT
20 PRINT "THIS PROGRAM CALCULATES THE LEVELIZED COST OF ENERGY FROM SOLAR ENERGY SYSTEMS, FROM VIEWPOINTS OF GOVERNMENT, HOMEOWNER, AND BUSINESS."
25 PRINT : PRINT "USER CAN CALCULATE LEVELIZED COST OF CONVENTIONAL ENERGY BY INPUTTING PRESENT VALUE OF ENERGY SAVINGS FROM PROGRAM PW SAVINGS INSTEAD OF CAPITAL COST."
27 PRINT : PRINT "WHAT ARE YOU LEVELIZING?": PRINT " 1) SYSTEM COSTS": PRINT "2) CONVENTIONAL ENERGY SAVINGS": INPUT XX: IF XX < 1 OR XX > 2 THEN 27
28 IF XX = 2 THEN PRINT : INPUT "NPW BENEFITS = ";CC: GOTO 100
30 PRINT : INPUT "CAPITAL COST = ";CC
40 INPUT "SALVAGE VALUE = ";S
50 INPUT "ANNUAL MAINTENANCE COSTS AS A PROPORTION OF CAPITAL COSTS = ";AM
60 INPUT "ANNUAL PROPERTY TAXES AS PROPORTION OF CAPITAL COSTS = ";PT
70 INPUT "FEDERAL RESIDENTIAL TAX CREDIT (AS DECIMAL) = ";FPC
80 INPUT "FEDERAL BUSINESS TAX CREDIT (AS DECIMAL) = ";FBC
90 INPUT "STATE TAX CREDIT (AS DECIMAL) = ";SC

95 INPUT "MARGINAL BUSINESS TAX RATE (AS DECIMAL)= ";T
 97 INPUT "MARGINAL PERSONAL TAX RATE (AS DECIMAL)= ";PE
100 INPUT "SYSTEM LIFETIME= ";Y
105 INPUT "DEPRECIABLE LIFE (ACCORDING TO ECONOMIC RECOVERY ACT). ENTER ZERO IF COMPUTING LEVELIZED BENEFITS. ";YD
110 INPUT "ANNUAL INTEREST RATE (AS DECIMAL)= ";I
120 INPUT "ARE ANNUAL ENERGY SAVINGS IN (1) KWH OR (2) MBTU (1/2)? ";X
130 IF X < 1 OR X > 2 THEN GOTO 120
140 IF X = 1 THEN X$ = "KWH": GOTO 160
150 X$ = "MBTU"
160 PRINT "ANNUAL ENERGY SAVINGS IN";X$;" = ";: INPUT ES
170 LD = (I * CC) / (1 − 1 / (I + 1) ^ Y): REM LEVELIZED CAPITAL COSTS
180 LPT = LD * (1 − FPC − (1 − PE) * SC) + CC * (AM + (1 − PE) * PT): REM COSTS WITH PERSONAL TAX CONSIDERATIONS, INCL. TAX DEDUCTIBILITY OF PROPERTY TAXES
185 IF YD = 0 THEN 200: REM SKIP DEPRECIATION SUBROUTINE IF CALCULATING LEVELIZED BENEFITS
190 GOSUB 320
200 LBT = LD * (1 − FBC − (1 − T) * SC) + CC * (1 − T) * (AM + PT) − DEP: REM COSTS WITH BUSINESS TAX DEDUCTIONS AND DEPRECIATION. N.B. FINANCING IGNORED BY ASSUMING EITHER 100% EQUITY OR 100% DEBTED AT STATED AFTER-TAX INTEREST RATE
205 IF YD = 0 THEN LLBT = (1 − T) * LBT: GOTO 220: REM REDUCE BENEFIT BY MARGINAL INCOME TAX RATE
210 LLBT = LBT/ (1 − T): REM REQUIRED REVENUES INCREASED TO PAY CORPORATE INCOME TAXES. DERIVATION FROM DOANE ET AL., APPENDIX E.
220 LG = LD + CC * AM: REM GOVERNMENTAL VIEWPOINT WITHOUT TAXES OR DEPRECIATION ALLOWANCES
225 PRINT : INPUT "PRINTER SLOT= (0=SCREEN)";SL: IF SL = 0 THEN 230
226 INPUT "NAME OF SITE= ";S$: INPUT "NAME OF SOLAR FILE= ";SF$: INPUT "SIZE OF SYSTEM=";SS$:D$ = CHR$ (4) : PRINT D$;" PR#";SL: PRINT "NAME OF SITE= ";S$: PRINT "NAME OF SOLAR FILE= ";SL$: PRINT "SIZE OF SYSTEM= ";SS$

230 PRINT : PRINT : PRINT "LEVELIZED SOCIAL CAPITAL COST = $"; INT (LD * 100 + .5) / 100
232 PRINT "LEVELIZED SOCIAL OPERATING COST = "; INT (CC * AM * 100 + .5) / 100
234 PRINT "TOTAL LEVELIZED SOCIAL COST = "; INT (LG * 100 + .5) / 100
236 PRINT "LEVELIZED HOMEOWNER COST = "; INT (LPT * 100 + .5) / 100
238 PRINT "LEVELIZED BUSINESS COST = "; INT (LBT * 100 + .5) / 100
240 PRINT "LEVELIZED SOCIAL COST PER ";X$;" = "; INT ((LG * 1000 + .5) / ES) / 1000
250 PRINT "POST—TAX LEVELIZED COST PER ";X$
260 PRINT " HOMEOWNER COST/";X$;" = "; INT ((LPT * 1000 + .5) / ES / 1000
270 PRINT " BUSINESS COST/";X$;" = "; INT ((LBT * 1000 + .5) / ES) / 1000
275 PRINT " BUSINESS REVENUE REQUIRED/";X$;" = "; INT ((LLBT * 1000 + .5) / ES) / 1000
280 PRINT : PRINT CHR$ (4) ;" PR# 0": INPUT "DO YOU WANT ANOTHER SCENARIO (Y/N)? ";Z$
290 IF LEFT$ (Z$,1) = "Y" THEN R = 0:P = 0: PRINT : GOTO 110
300 END
310 REM WRITTEN BY MARTIN KATZMAN MAY 1982, CURRENT VERSION AUG. 1983.
320 REM SUBROUTINE DEPRECIATION
330 D = CC − S: REM AMOUNT TO DEPRECIATE
350 PRINT "ENTER DEPRECIATION METHOD:"
360 PRINT "(1) SUM-OF-YEARS-DIGITS"
370 PRINT " (2) DECLINING BALANCE"
375 PRINT " (3) STRAIGHT—LINE"
380 INPUT X
390 IF X < 1 OR X > 3 THEN 350
400 ON X GOTO 410,550,660
410 REM SUM-OF-YEARS-DIGITS-METHOD
420 REM ACDEP IS THE ACCUMULATED DEPRECIATION. PVDEP IS PRESENT VALUE OF DEPRECIATION ALLOWANCE.
430 ACDEP = 0
440 FOR J = 1 TO YD
450 R = 2 * D * (YD − J + 1) / ((YD + 1) * YD)
460 P = P + (R * T) / (1 + I) ^ J
470 ACDEP = ACDEP + R

480 BAL = D − ACDEP: REM B IS REMAINING BALANCE TO DEPRECIATE
490 IF BAL > = 0 THEN 530: REM TEST FOR REMAINING BALANCE
500 ACDEP = ACDEP + BAL
510 R = R + BAL
520 BAL = 0
530 NEXT J
540 GOTO 700
550 REM DECLINING BALANCE METHOD
560 INPUT "% DEPRECIATION ";M
570 M = M / 100: REM CONVERT PERCENT TO DECIMAL
580 R = D
590 FOR J = 1 TO YD
600 ACDEP = R * M / YD
610 P = P + (ACDEP * T) / (1 + I) ^ J
620 REM − ACCUMULATE REMAINING BALANCE
630 R = R − ACDEP
640 NEXT J
650 GOTO 700
660 REM STRAIGHT-LINE DEPRECIATION
670 FOR J = 1 TO YD
680 P = P + (CC * T / YD) / (1 + I) ^ J: NEXT
700 DEP = (I * P) / (1 − 1 / (I + 1) ^ Y): REM LEVELIZED DEPRECIATION, OVER ECONOMIC LIFETIME OF SYSTEM, NOT ACCOUNTING LIFETIME
710 RETURN

Program PW SAVINGS

10 TEXT : HOME : PRINT "PROGRAM PW SAVINGS:": PRINT
20 PRINT "THIS PROGRAM CALCULATES THE PRESENT WORTH OF FUEL SAVINGS AND MAINTENANCE COSTS AND NET PRESENT WORTH OF SOLAR ENERGY SYSTEMS.": PRINT
30 INPUT "CAPITAL COST OF SYSTEM = ";CC
40 INPUT "ANNUAL MAINTENANCE AS PROPORTION OF CAPITAL COSTS ";AM:AM = AM * CC
50 INPUT "SYSTEM LIFETIME = ";YR
60 INPUT "ANNUAL ENERGY SAVED IN MBTU OR KWH ";ES
70 INPUT "VALUE OF ENERGY PER MBTU OR PER KWH ";VS:EV = ES * VS

80 INPUT "INTEREST RATE (AS PERCENT)= ";I1:I = I1 / 100
 90 INPUT "ENERGY ESCALATION RATE (AS PERCENT)= ";E1:ER = E1 / 100
100 SVE = 0:SE = 0:SM =0
110 FOR J = 1 TO YR
120 SVE = SVE + EV * ((1 + ER) / (1 + I)) ^ J
130 SE = SE + EV / (1 + I) ^ J
140 SM = SM + AM / (1 + I) ^ J
150 NEXT
160 INPUT "SLOT NUMBER (0 = SCREEN) ";SL:D$ = CHR$ (4): PRINT D$;"PR# ";SL
170 Sl = INT (SE + .5):S2 = INT (SVE + .5):S3 = INT (SM + .5):B1 = S1 + S3:B2 = S2 − S3:N1 = B1 − CC:N2 = B2 − CC
180 HOME : PRINT : PRINT " PRESENT VALUE OF ENERGY SAVINGS": PRINT : PRINT TAB(27);"ANNUAL FUEL": PRINT TAB(25);"ESCALATION RATE": PRINT TAB(25): FOR I = 1 TO 3: PRINT "= = = = =";: NEXT : PRINT
190 PRINT TAB(27);"0%"; TAB(37);E1;"%": PRINT TAB(26) ; "= = = ="; TAB(36);"= = = ="
200 A1 = LEN (STR$ (S1)):A2 = LEN (STR$ (S2)):A3 = LEN (STR$ (S3)):A4 = LEN (STR$ (B1)):A5 = LEN (STR$ (B2)):A6 = LEN (STR$ (N1)):A7 = LEN (STR$ (N2))
210 PRINT "PW FUEL"; TAB(20); SPC(10 − A1) ;S1; SPC(10 − A2) ;S2
220 PRINT "PW MAINT"; TAB(20); SPC(10 − A3) ;S3; SPC(10 − A3) ;S3
230 FOR I = 1 TO 10: PRINT "----";: NEXT : PRINT
240 PRINT " NET BENEFITS"; TAB(20); SPC(10 − A4) ;B1; SPC(10 − A5) ;B2
250 PRINT " CAPITAL COSTS"; TAB(20); SPC(10 − LEN (STR$ (CC)));CC; TAB(10); SPC(10 − LEN (STR$ (CC)));CC
260 FOR I = 1 TO 8: PRINT "= = = = =";: NEXT
270 PRINT "NET PRESENT WORTH"; TAB(20); SPC(10 − A6); N1;SPC(10 − A7) ;N2
280 PRINT : PRINT
290 PRINT D$;"PR# 0"
300 INPUT "DO YOU WANT TO ANALYZE ANOTHER SITE? (Y/N)";X$
310 IF LEFT$ (X$,1) = "Y" THEN PRINT : GOTO 60
320 REM WRITTEN BY MARTIN KATZMAN IN APPLESOFT BASIC, MAY 1982. REVISED IN JULY 1982.

Program WINDMILL PERFORMANCE (abstract)

****

WINDMILL PERFORMANCE

WRITTEN BY
ARLENE AND MARTIN KATZMAN

UNIVERSITY OF TEXAS AT DALLAS

COPYRIGHT OCTOBER 1983

****

THIS PROGRAM PERFORMS THE FOLLOWING CALCULATIONS FOR UP TO 10 WECS:

- FITS WIND SPEED-WIND POWER CURVE
- PRODUCES A WIND POWER DISK FILE
- CALCULATES ANNUAL KWH GENERATED, ENERGY SOLD BACK, AND LOAD REDUCTION
- COMPUTES ANNUAL VALUE OF ENERGY

USER INPUTS HOURLY WIND SPEED AND LOAD FILES. OWN LOAD CAN BE REPRESENTED BY SCALED-DOWN UTILITY LOAD.

Program GENCOST (abstract)

****

GENCOST

WRITTEN BY
ARLENE, DOUGLAS, AND MARTIN KATZMAN

UNIVERSITY OF TEXAS AT DALLAS

COPYRIGHT OCTOBER 1983

****

GENCOST IS A PRODUCTION COSTING MODEL. GENCOST COMPUTES ELECTRIC GENERATION BY EACH FUEL TYPE AT SEVERAL LEVELS OF SOLAR OR WIND PENETRATION. RESULTS MAY BE INPUT TO KILOCOST.

USER INPUTS UP TO 9 UNIT TYPES; E.G., NUCLEAR, COAL. MAINTENANCE SHOULD BE SCHEDULED IN TROUGHS OF WEEKLY PEAK LOADS AS INDICATED BY WEEKPEAK GRAPHS. IF UNIT IS NOT MAINTAINED THEN ENTER 53,54 AS STARTING, ENDING WEEKS OF MAINTENANCE.

GENCOST CAN SCALE SOLAR PENETRATION AS 1) ABSOLUTE MW; 2) PROPORTION OF PEAK LOAD; OR 3) AS PERCENT OF HOUSEHOLDS WITH 5KW SYSTEMS. GENCOST CAN SCALE WIND PENETRATION BY (1) AND (2) ONLY.

Program LOLP (abstract)

****

LOLPS

WRITTEN BY
ARLENE AND MARTIN KATZMAN

UNIVERSITY OF TEXAS AT DALLAS

COPYRIGHT OCTOBER 1983

****

LOLPS CALCULATES LOSS OF LOAD PROBABILITIES AT SEVERAL LEVELS OF SOLAR PENETRATION. RESULTS CAN BE USED TO COMPUTE CAPACITY SAVINGS.

LOLPS CAN SCALE SOLAR PENETRATION AS 1) ABSOLUTE MW; 2) PROPORTION OF PEAK LOAD; OR 3) AS PERCENT OF HOUSEHOLDS WITH 5KW SYSTEMS.

LOLPS REQUIRES 3 UNIT TYPES, GROUPED BY SIZE AND FORCED OUTAGE RATE, E.G., UNITS OF ABOUT 100MW AND .05 FORCED OUTAGE. USER ALLOWED UP TO 15 UNITS OF EACH TYPE. PROGRAM WORKS FASTER IF UNIT SIZES ARE MULTIPLES, E.G. 1000-500-250MW. USER SCHEDULES MAINTENANCE FROM WEEKPEAK.

Bibliography

Abelson, John. 1980. "A revolution in biology." *Science* 209 (September):1319–21.
Abelson, Philip H., ed. 1974. *Energy: Use, Conservation, and Supply*. Washington, D.C.: American Association for the Advancement of Science.
Abelson, Philip H., and Hammond, Allen L. 1978. *Energy II: Use, Conservation, and Supply*. Washington, D.C.: American Association for the Advancement of Science.
Adler, David. 1980. "Amorphous silicon offers promise of cost reductions." *Solar Engineering* 5 (September):39–40.
Arrow, Kenneth. 1962. "The economic implications of learning by doing." *Review of Economic Studies* (June):155–73.
Asbury, Joseph G., and Kouvalis, A. 1976. *Electric Storage Heating: The Experience in England and Wales and Federal Republic of Germany*. ANL/ES–50. Argonne National Laboratory (May).
Asbury, Joseph, and Mueller, Ronald O. 1977. "Solar energy and electric utilities: Should they be interfaced?" *Science* 195 (February): 445–50.
Ascher, William. 1978. *Forecasting: An Appraisal for Policy-Makers and Planners*. Baltimore: Johns Hopkins University Press.
Avery, John G., and Krall, John J. 1982. "Corrosion prevention and fluid maintenance in active solar systems: The state-of-the art." *Solar Engineering and Contracting* 1 (February):22–26; ibid. (March):30–36; ibid. (April):22–26.
Ayer, James. 1981. "How utilities can prepare for the advent of solar energy." *Electric Light & Power* 59 (March):54–55.
Ayres, Robert U. 1968. "Envelope curve forecasting." Pp. 77–94 in James R. Bright, ed., *Technological Forecasting for Industry and Government*. Englewood Cliffs, N.J.: Prentice-Hall.
Baughman, Martin L.; Joskow, Paul L.; and Kamat, Dilip P. 1979. *Electric Power in the United States: Models and Policy Analysis*. Cambridge, Mass.: MIT Press.
Berg. Charles A. 1978. "Process innovation and changes in industrial use." Pp. 3–9 in Abelson and Hammond.

Berndt, Ernst R., and Wood, David O. 1979. "Engineering and econometric interpretations of energy-capital complementarity." *American Economic Review* 69 (September):342–54.
Bickler, Donald B.; Gallagher, Brian D.; and Sanchez, Lloyd E. 1978. "A candidate low-cost processing sequence for terrestrial silicon solar cell panel." Pp. 241–45 in *Proceedings of 13th IEEE Specialists Conference.*
Binswanger, Hans, et al. 1978. *Induced Innovation.* Baltimore: Johns Hopkins University Press.
Blankle, Gerald. 1981. "Co-generation poses difficult distribution system problems." *Electric Light & Power* 59 (August):60–63.
Blythe, Paul, Jr. 1981. "Thin-film solar cell research progress." *Solar Engineering* 6 (April):12–17.
Bohm, Robert A.; Clinard, Lillian A.; and English, Mary R. 1981. *Energy Production and Productivity.* Proceedings of the International Energy Symposium. Vol. I. Cambridge, Mass.: Ballinger Publishing Company.
Bolton, James R., and Hall, David O. 1979. "Photochemical conversion and storage of solar energy." *Annual Review of Energy* 4 :353–401.
Breneman, W. C.; Farrier, E. G.; and Morihara, H. 1978. "Preliminary process design and economics of low-cost solar-grade silicon production." Pp. 339–43 in *Proceedings of 13th IEEE Specialists Conference.*
Bright, James R. 1973. "The process of technological innovation: An aid to understanding technological forecasting." Pp. 3–12 in Bright and Schoeman.
Bright, James R., and Schoeman, Milton E. F., eds. 1973. *A Guide to Practical Technological Forecasting.* Englewood Cliffs, N.J.: Prentice-Hall.
Brinkworth, B. J. 1977. "Autocorrelation and stochastic modeling of insolation sequences." *Solar Energy* 19: 343–47.
Brown, Kenneth C. 1981. "Solar industrial process heat systems: Cost and performance vary widely." *Solar Engineering* 6 :22–34.
Brown, Lester R. 1980. "Food or Fuel: New Competition for the World's Cropland." *World Watch Paper No. 35* (March) Washington, D.C.: World Watch Institute.
———. 1981. *Building a Sustainable Society.* New York: W.W. Norton.
Burt Hill Kosar Rittelman Associates. 1979. *Residential Photovoltaic Module and Array Requirement Study.* DOE/JPL–955149–79/1. Prepared for the Jet Propulsion Laboratory.
Burwell, C. C. 1978. "Solar biomass energy: An overview of U.S. potential." Pp. 161–68 in Abelson and Hammond.
Butti, Ken, and Perlin, John. 1980. *A Golden Thread: 2500 Years of Solar Architecture and Technology.* Palo Alto, Calif.: Cheshire Books. Chaps. 1–3.
Calvin, Melvin. 1974. "Solar energy by photosynthesis." Pp. 111–17 in Abelson.
Cassiday, Bruce N. 1978. *The Complete Solar House.* New York: Warner Communications.
Castle, Emery, and Hoch, Irving. 1982. "Farm real estate price components, 1920–78." *American Journal of Agricultural Economics* 64 (February): 8–18.
Cheng, Edmond, and Wong, Benedict. 1979. "Stochastic simulation of hourly surface wind in Hawaii." University of Hawaii, College of Engineering (April).
Chopra, Prem S. 1980. "Why do solar systems fail?" *Solar Engineering* 5 (October):14–21.
CONAES (Committee on Nuclear and Alternative Energy Systems) National Research Council. 1979. *Energy in Transition, 1985–2010, Final Report.* San Francisco: W.H. Freeman for the National Academy of Sciences.
CONOCO, Inc., Coordinating and Planning Dept. 1982. *World Energy Outlook through 1990* (January).
Consumer's Union. 1982. "Solar water heaters." *Consumer Reports* 47 (May): 256–61.
Corey, Gordon R. 1983. "Plant investment decision making in the electric power industry." Pp. 377–403 in Lind.
Costello, Dennis, et al. 1978. *Photovoltaic Venture Analysis, Final Report.* SERI/TR-52-040. Vol. I. Solar Energy Research Institute (July).
Crew, Michael A., and Kleindorfer, Paul R. 1979. *Public Utility Economics.* New York: St. Martin's Press.

Daniels, Farrington. 1977. *Direct Use of the Sun's Energy.* New York: Ballantine. Reprinted from 1954 edition.
Deese, David A., and Nye, Joseph S., eds. 1981. *Energy and Security.* Cambridge, Mass.: Ballinger Publishing Co.
de Marchi, Neil. 1981. "The Ford Administration: Energy as a political good." Pp. 475–546 in Craufurd D. Goodwin, ed. *Energy Policy in Perspective: Today's Problems, Yesterday's Solutions.* Washington, D.C.: Brookings Institution.
Doane, James W., et al. 1976. *The Cost of Energy from Utility-Owned Electric Systems: A Required Revenue Methodology for ERDA/EPRI Evaluations.* JPL 5040-29. Jet Propulsion Laboratory (June).
Dvoskin, D., and Heady, Earl O. 1976. *Energy Requirements of Irrigated Crops in the Western United States.* Miscellaneous Report, Iowa State University, Center for Agricultural and Rural Development (November).
EEI (Edison Electric Institute). 1976. *Economic Growth in the Future: The Growth Debate in National and Global Perspectives.* New York: McGraw-Hill.
———. 1974. *Historical Statistics of the Electric Utility Industry through 1970.* Publication No. 73–74 (April).
———. 1979. *Statistical Yearbook of the Electric Utility Industry for 1978.* No.46 (November).
———. 1980. *Statistical Yearbook of the Electric Utility Industry/1979.* No.47 (November).
Eldridge, Frank R. 1980. *Wind Machines.* New York: Van Nostrand Reinhold.
Endrenyi, J. 1978. *Reliability Modeling in Electric Power Systems.* New York: John Wiley & Sons.
EPRI. 1977. *Synthetic Electric Utility Systems for Evaluating Advanced Technologies.* EPRI EM-285. Electric Power Research Institute (February).
Everitt, Neil. 1981. "Off-peak storage heating: The British experience." *Electric Light & Power* 59 (July): 56–58.
Exxon Corporation, Corporate Planning and Public Affairs Dept. 1982. *World Energy Outlook,* Background Series (December).
Feldman, Stephen D., and Wirtshafter, Robert M. 1980. *On the Economics of Solar Energy: The Public-Utility Interface.* Lexington, Mass.: D.C. Heath.
Fenn, Scott. 1981. *The Nuclear Power Debate: Issues and Choices.* New York: Praeger.
Ferraro, A. G. 1969. "Valuation of property interests for ad valorem taxation of extractive industry and agricultural realty." In Arthur D. Lynn, ed., *The Property Tax and Its Administration.* Madison: University of Wisconsin Press.
Finger, Susan. 1979. *Electric Power System Production Costing and Reliability Analysis including Hydroelectric, Storage, and Time Dependent Power Plants.* MIT-EL-79-006. Cambridge: MIT Energy Laboratory (February).
———. 1980. *SYSGEN Production Costing and Reliability Model User Documentation.* MIT-EL-79-020. Cambridge: MIT Energy Laboratory Technical Report (June).
Fisher, Anthony C. 1979. "Measures of natural resource scarcity." Pp. 249–75 in V. Kerry Smith, ed. *Growth and Scarcity Revisited.* Baltimore: Johns Hopkins University Press.
Fusfeld, Alan R. 1973. "The technological progress function: A new technique for forecasting." Pp. 92-105 in Bright and Schoeman.
Gay, Charles F. 1980. "Solar cell technology: An assessment of the state of the art." *Solar Engineering* 5 (March): 15–18.
General Electric. 1978. *Requirements Assessment of Photovoltaic Power Plants in Electric Utility Systems.* EPRI ER-685-ST. 3 vols. Prepared for Electric Power Research Institute (June).
———. 1979. *Regional Conceptual Design and Analysis Studies for Residential Photovoltaic Systems.* SAND78-7039. 2 vols. Prepared for Sandia Laboratories (January).
Giese, R. F. 1979. *SIMSTOR, A Cost Allocation Model for Assessing Electric Heating and Cooling Technologies.* ANL/SPG-5. Argonne National Laboratory (May).
Goldemberg, Jose. 1978. "Brazil: Energy options and current outlook." Pp. 28–34 in Abelson and Hammond.
———. 1981. "A centralized 'soft' energy path." Pp. 187–99 in Bohm, Clinard, and English.
Golding, E. W. 1976. *The Generation of Electricity by Wind Power.* 2nd ed. New York: John Wiley & Sons/Halsted Press.

Goodwin, Craufurd D., ed. 1981. *Energy Policy in Perspective: Today's Problems, Yesterday's Solutions.* Washington, D.C.: Brookings Institution.

Green, Donald E. 1973. *The Land of the Underground Rain.* Austin: University of Texas Press.

Grenon, L. A., and Coleman, M. G. 1978. "Silicon solar cells, a manufacturing cost analysis." Pp. 246–251 in *Proceedings of 13th IEEE Specialists Conference.*

Griliches, Zvi. 1960. "Hybrid corn and the economics of innovation." *Science* 132 (July): 275–80.

Haefele, Wolf, et al. 1981. *Energy in a Finite World.* 2 vols. Report of the Energy Systems Program Group of the International Institute for Applied Systems Analysis. Cambridge, Mass.: Ballinger Publishing Co.

Hamilton, R. C., and Witt, C. E. 1978. "Terrestrial concentrating photovoltaic systems comparative performance and costs." Pp. 920–24 in *Proceedings of 13th IEEE Specialists Conference.*

Hansen, J., et al. 1981. "Climate impact of increasing atmospheric carbon dioxide." *Science* 213 (August): 957–66.

Harlan, James K. 1982. *Starting with Synfuels.* Cambridge, Mass.: Ballinger Publishing Co.

Hausman, Jerry A. 1979. "Individual discount rates and the purchase and utilization of energy-using durables." *Bell Journal of Economics* 10: 33–54.

Haveman, Robert H. 1969. "The opportunity cost of displaced private spending and the social discount rate." *Water Resources Research* 5 (October): 947–57.

Heal, Geoffrey. 1976. "The relationship between price and extraction cost for a resource with a backup technology." *Bell Journal of Economics* 7: 371–78.

Hein, Gerald F.; Cusick, James P.; and Poley, William A. 1978. "Impact of balance of system (BOS) costs on photovoltaic power systems." Pp. 930–33 in *Proceedings of 13th IEEE Specialists Conference.*

Hirst, Eric, and Carney, Janet. 1977. *Residential Energy Use to the Year 2000: Conservation and Economics.* Oak Ridge National Laboratory, ORNL/CON (September).

Hoyle, Fred, and Hoyle, Geoffrey. 1980. *Commonsense in Nuclear Energy.* San Francisco: W. H. Freeman.

Hudson, Edward A., and Jorgenson, Dale W. 1974. "U.S. energy policy and economic growth, 1975–2000." *Bell Journal of Economics* 5: 461–514.

Hunt, L. P., et al. 1978. "Advances in the Dow Corning process for solar grade silicon." Pp. 333–38 in *Proceedings of 13th IEEE Specialists Conference.*

Inglis, David R. 1978. *Wind Power and Other Energy Options.* Ann Arbor: University of Michigan Press.

Jacobsen, A. S., and Ackerman, P. D. 1981. "Cost reduction projections for active solar systems. Pp. 1291–93 in *Proceedings of 1981 Annual Meeting.* Philadelphia: American Section/International Solar Energy Society, (May).

Javetski, John. 1979. "A burst of energy in photovoltaics: A special report." *Electronics* (July): 105–22.

Jones, Byron W. 1982. *Inflation in Engineering Economic Analysis.* New York: John Wiley & Sons.

Jorgenson, Dale W. 1981. "Energy prices and productivity growth." *Scandinavian Journal of Economics* 83, no. 2: 165–79.

Kammer, W. A. 1979. "A preliminary analysis of the impact on utility distribution systems of power feedback from residential photovoltaic systems." ATR-79 (7694-07)-2. Aerospace Corporation (March).

Katzman, Martin T. 1981. "Paradoxes in the diffusion of a rapidly advancing technology: The case of solar photovoltaics." *Technological Forecasting and Social Change* 19: 227–36.

———. 1983. "Economic prospects for wind farming: A simulation approach." *Energy Exploration and Exploitation* 2, no. 1: 67–80.

Katzman, Martin T., and Katzman, Arlene C. 1982. "Simulating impacts of photovoltaics on utilities." Pp. 1109–14 in *Progress in Solar Energy.* Proceedings of the 1982 meetings of the American Solar Energy Society.

———. 1983. "Electricity production costing: The validity of micro-computer simulation." *Electric Ratemaking* 2 (April/May): 37–40.

Katzman, Martin T., and Matlin, Ronald W. 1978. "The economics of adopting solar energy systems for crop irrigation." *American Journal of Agricultural Economics* 60 (November): 648–54.
Kelly, Henry. 1978. "Photovoltaic power systems: A tour through the alternatives." Pp. 151–60 in Abelson and Hammond.
Keran, Michael W. 1976. "Inflation, regulation, and utility stock prices." *Bell Journal of Economics* 7: 268–80.
Klass, D. L. 1978. "Anaerobic digestion of macrocystis pyrifera under mesophilic conditions." *Symposium Papers*. Chicago: Institute of Gas Technology (March).
Klebba, James M. 1980. "Insuring solar access on retrofits: The problem and some solutions." *Solar Engineering* 5 (January): 16–19.
Kran, Alexander. 1978. "Silicon ribbon technology assessment, 1978-1986: A computer-assisted analysis using PECAN." Pp. 344–49 in *Proceedings of 13th IEEE Specialists Conference*.
Krenz, Jerrold H. 1976. *Energy: Conversion and Utilization*. Boston: Allyn & Bacon.
Kuznets, Simon. 1966. *Modern Economic Growth*. New Haven: Yale University Press.
Laitos, Jan, and Feuerstein, Randall. 1979. *Regulated Utilities and Solar Energy*. SERI/TR-62-255. Solar Energy Research Institute (June).
Landsberg, Hans, et al. 1979. *Energy: The Next Twenty Years*. Report of a Study Group Sponsored by the Ford Foundation. Cambridge, Mass.: Ballinger Publishing Co.
Lane, D. E.; Fischbach, P. E.; and Teter, N. C. 1973. *Energy Use in Nebraska Agriculture*. CC255. University of Nebraska–Lincoln, College of Agriculture.
Lenz, Ralph, C., Jr. 1968. "Forecasts of exploding technologies by trend extrapolation." Pp. 57–76 in James C. Bright, ed. *Technological Forecasting for Industry and Government*. Englewood Cliffs, New Jersey: Prentice-Hall.
Lind, Robert C., ed. 1983. *Discounting for Time and Risk in Energy Policy*. Baltimore: Johns Hopkins University Press.
———. 1983a. "A primer on the major issues relating to the discount rate for evaluating national energy options." Pp. 21–94 in Lind.
———. 1983b. "The rate of discount and the application of social-benefit cost analysis in the context of energy policy decisions." Pp. 443–58 in Lind.
Lonnroth, Mans; Steen, Peter; and Johansson, Thomas B. 1980. *Energy in Transition: A Report on Energy Policy and Future Options*. Berkeley and Los Angeles: University of California Press.
Lovins, Amory B. 1977. *Soft Energy Paths: Toward a Durable Peace*. Harmondsworth: Penguin Books.
Lovins, Amory B, and Lovins, L. Hunter. 1982. *Brittle Power: Energy Strategy for National Security*. Andover, Mass.: Brick House.
Mackintosh, B. H., et al. 1978. "Multiple silicon ribbon growth by EFG." Pp. 350–57 in *Proceedings of 13th IEEE Specialists Conference*.
Mansfield, Edwin. 1971. *Technological Change*. New York: W. W. Norton.
Marsh, W. D. 1980. *Economics of Electric Utility Power Generation*. Oxford Engineering Science Series. Oxford: Oxford University Press.
Matlin, Ronald W., and Katzman, Martin T. 1977. *The Economics of Adopting Solar Photovoltaic Energy Systems in Irrigation*. COO/4094-2. MIT Lincoln Laboratory (December).
———. 1979. "Assessing solar photovoltaic energy systems for crop irrigation." *Water Resources Bulletin* 15 (October): 1308–17.
Matlock, John H. 1979. "Crystal growing spotlight." *Semiconductor International*. (October): 33–44.
Maycock, Paul D. 1978. "The development of photovoltaics as a power source of large-scale terrestrial application." Pg. 6 in *Proceedings of 13th IEEE Photovoltaics Specialists Conference*.
McGraw, Michael G. 1981. "Wind turbines: An idea whose time has come—again." *Electrical World* (May): 97–110.
Metz, William D., and Hammond, Allen D. 1978. *Solar Energy in America*. Washington, D.C.: American Academy for the Advancement of Science.
Moore, Robert H. 1976. "Cost prediction for PV sources." *Solar Energy* 18, no. 3: 225–34.

Mueller, R. O.; Cha, B. K.; and Giese, R. F. 1981. "Solar photovoltaic power systems: Will they reduce utility peaking requirements?" *Science* 213 (July): 211–13.

Murray, William J. 1981. "PV in 1980: Growth accelerates." *Solar Engineering* 6 (January): 14–16.

National Research Council. 1982. "Fuel science and technology." Chap. 15 in *Outlook for Science and Technology: The Next Five Years.* A Report of the National Academy of Sciences. San Francisco: W. H. Freeman & Co.

———. 1983. *Acid Deposition in Eastern North America.* Washington, D.C.: National Academy Press.

Neff, Thomas L. 1981. *The Social Costs of Solar Energy: A Study of Photovoltaic Energy Systems.* New York: Pergamon Press.

Nelkin, Dorothy, and Pollak, Michael. 1981. *The Atom Besieged: Extra-Parliamentary Dissent in France and Germany.* Cambridge: MIT Press.

NERC 1982a. *Twelfth Annual Review of Overall Reliability and Adequacy of North American Bulk Power Systems.* Princeton, N.J.

———. 1982b. *1982 Annual Report.* Princeton, N.J.

———. 1983. *Electric Power Supply and Demand, 1983–1992, for Regional Reliability Councils of NERC.* Princeton, N.J.

Nesbit, William. 1979. *World Energy: Will There Be Enough in 2020.* Decisionmaker Bookshelf. Vol. 6. Washington, D.C.: Edison Electric Institute.

Nordhaus, William D. 1979. *The Efficient Use of Energy Resources.* Cowles Foundation Monograph 26. New Haven: Yale University Press.

North, W. J. 1977. "Growth factors in the production of giant kelp." In *Symposium Papers: Clean Fuels from Biomass and Waste.* Chicago: Institute of Gas Technology (March).

Office of Technology Assessment (OTA), U.S. Congress. 1980. *Residential Energy Conservation.* Montclair, N.J.: Allanheld, Osmun & Co.

OECD, International Energy Agency. 1982. *World Energy Outlook.* Paris: Organization for Economic Cooperation and Development.

Palson, B. O., et al. 1981. "Biomass as a source of chemical feedstocks: An economic evaluation." *Science* 213 (July): 513–17.

Palz, Wolfgang. 1978. *Solar Electricity: An Economic Approach to Solar Energy.* London: Butterworths.

Peck, Merton J., and Scherer, Frederic. 1962. *The Weapons Acquisition Process.* Cambridge: Harvard University Press.

Perino, Audrey M. 1979. *A Methodology for Determining the Economic Feasibiity of Residential or Commercial Solar Energy Systems.* SAND78-0931. Sandia National Laboratory (January).

Perl, Lewis J., and Dunbar, Frederick C. 1982, "Cost effectiveness and cost-benefit analysis of air quality regulations." *American Economic Review* 72 (May): 208–13.

Pettway, Richard H. 1978. "On the use of beta in regulatory proceedings: An empirical examination." *Bell Journal of Economics* 9: 239–48.

Pimentel, David, et al. 1968. "Food production and the energy crisis." Pp. 41–47 in Abelson.

———. 1981. "Biomass energy from crop and forest residues." *Science* 212, no. 5 (June): 1110–15.

Plummer, James L. 1981. "Methods for measuring the oil import reduction premium and the oil stockpile premium." *Energy Journal* 2 (January): 1–18.

Prince, Morton. 1963. "Latest developments in the field." *Solar energy* 7, no. 4: 185.

Ramsay, William. 1979. *Unpaid Costs of Electrical Energy: Health and Environmental Impacts from Coal and Nuclear Power.* Baltimore: Johns Hopkins University Press for Resources for the Future.

Rapp, Donald. 1981. *Solar Energy.* Englewood Cliffs, N.J.: Prentice-Hall.

Roberts, D. R. 1981. "Westinghouse aims at high efficiency, low cost system." *Solar Engineering* 6 (August): 12–15.

Roessner, R., et al. 1979. *Application of Diffusion Research to Solar Energy Policy Issues.* SERI/TR-51-194. Solar Energy Research Institute (March).

Roll, Richard. 1972. "Interest rates on monetary assets and commodity price index changes." *Journal of Finance* 27 (May): 251–77.

Rothman, Harry; Greenshields, Rod; and Calles, Francisco Rosillo. 1983. *Energy from Alcohol: The Brazilian Experience.* Lexington: University Press of Kentucky.
Rowen, Henry S., and Weyant, John P. 1982. "Reducing the economic impacts of oil supply interruptions: An international perspective." *Energy Journal* 3 (January): 1–34.
Roy, Aharon. 1978. "Model for comparing cost of flat-array and concentrator photovoltaic solar-cell systems." Pp. 914–19 in *Proceedings of 13th IEEE Specialists Conference.*
Russell, Miles C. 1979. *Solar Photovoltaic/Thermal Residential Systems.* C00-4577-9. MIT Lincoln Laboratory (December).
Ryther, John H. 1979. "Biomass production by marine and freshwater plants." *Proceedings of Third Annual Energy Systems Conference.* National Biomass Program. Golden, Colo.: Solar Energy Research Institute.
Sargent, Thomas J. 1973. "Interest rates and prices in the long run." *Journal of Money, Credit and Banking* 5 (February): 395–449.
Schiffel, D. 1978 *Measuring the Potential Benefits of Government-Accelerated Technological Innovation: The Case of Photovoltaics.* SERI/DDS-040. Solar Energy Research Institute.
Schmidt, Phillip S. 1983. *Electricity and Industrial Productivity: A Technical and Economic Perspective.* Electric Power Research Institute.
Schmookler, Jacob. 1966. *Invention and Economic Growth.* Cambridge: Harvard University Press.
Schurr, Sam H., et al. 1979. *Energy in America's Future: The Choices Before Us.* Baltimore: Johns Hopkins University Press for Resources for the Future.
Senior, Thomas B., and Sengupta, Dipak L. 1981. "An environmental effect of large wind turbines." *Solar Engineering* 6 (August): 22–25.
Shapira, Hanna; Brite, Stephen E.; and Yost, Michael B. 1981. "Up and down: Energy and cost comparisons." Pp. 931–35 in *Proceedings of 1981 Annual Meeting.* Philadelphia: American Section/International Solar Energy Society (May).
Sheridan, Norman R. 1972. "Criteria for justification of solar energy systems." *Solar Energy* 13: 426.
Sloggett, Gordon. 1977. *Energy and U.S. Agriculture: Irrigation and Pumping, 1974.* Agricultural Economic Report No. 376. U.S. Dept. of Agriculture, Economic Research Service.
Smil, Vaclav. 1981. "Energy in rural China." Pp. 309–28 in Clinard, English, and Bohm.
Smock, Robert W. 1981a. "Coal: Utilities fuel of the future." *Electric Light & Power* 59 (March): 25–30.
———. 1981b. "Success elusive in garbage power." *Electric Light & Power* 59 (July): 25–28; and ibid. (August): 68–74.
———. 1981c. "The top 100 electric utilities 1980 operating performance." *Electric Light & Power* 59 (August): 17–20.
Snell, Jack E.; Achenbach, Paul R.; and Petersen, Stephen R. 1968. "Energy conservation in new housing design." Pp. 86–92 in Abelson.
Solow, Robert M., and Wan, Frederic Y. 1976. "Extraction costs in the theory of exhaustible resources." *Bell Journal of Economics* 7: 359–70.
Stauffer, Thomas. 1983. "The social efficiency of electric utility decision criteria." Pp. 412–42 in Lind.
Stelzer, Irwin. 1980. "The electric utilities face the next twenty years." Pp. 127–44 in Hans Landsberg, ed. *Selected Studies on Energy: Background Papers for Energy: The Next Twenty Years.* Cambridge, Mass.: Ballinger Publishing Co.
Stirewalt, Edward N. 1981. "Get set for solar electric cars." *Solar Age* 6 (September): 22–27.
Tabors, Richard D.; Finger, Susan; and Cox, Alan J. 1981. "Economic operation of distributed power systems within an electric utility." *IEEE Transactions on Power Apparatus and Systems.* Vol. PAS-100. No. 9 (September).
Temin, Peter, 1966. "Steam and waterpower in the early 19th century." *Journal of Economic History* (June): 187–205.
Thompson, Howard E. 1979. "Estimating the cost of equity capital for electric utilities: 1958–1976." *Bell Journal of Economics* 10: 619–38.
Ul-Rahman, Saif M. 1967. "Prospects and limitations of solar energy utilization in developing countries." *Solar Energy* 11: 102.

U.S. Department of Energy. 1979. *National Photovoltaic Program: Multi-Year Program Plan.* Draft June 6, 1979.

U.S. Department of Housing and Urban Development (HUD). 1979. *State Solar Legislation as of January 1979.*

Uri, N. D., and Hassanen. 1982. "Energy prices, labor productivity, and causality: An empirical examination." *Energy Economics* 4 (April): 98–104.

Vardi, Joseph, and Avi-Itzhak, Benjamin. 1981. *Electric Energy Generation: Economics, Reliability, Rates.* Cambridge: MIT Press.

Wade, Nicholas. 1968. "Windmills: The resurrection of an ancient energy technology." Pp. 128–30 in Abelson.

Weisz, Paul B., and Marshall, John F. 1979. "High-grade fuels from biomass farming: Potentials and constraints." *Science* 206 (October).

Wenders, John T. 1976. "Peak load pricing in the electric utility industry." *Bell Journal of Economics* 7: 234–41.

Westinghouse R & D Center. 1979. *Regional Conceptual Design and Analysis for Residential Photovoltaic Systems. Vol. I. Executive Summary.* SAND78-7040. Prepared for Sandia National Laboratory (September).

White, Sharon. 1979. *Municipal Bond Financing of Solar Energy Cacilities.* SERI/TR-62-191. Solar Energy Research Institute (July).

Wilkening, Harold A. 1978. "Design of a 10 kw photovoltaic 200/1 concentrator." Pp. 669–72 in *Proceedings of 13th IEEE Specialist Conference.*

Wilson, Carroll L. 1980. *Coal: Bridge to the Future.* Report of the World Coal Study. Cambridge, Mass.: Ballinger Publishing Co.

Wolf, Martin; Goldman, Howard M.; and Lawson, Albert C. 1978. "Evaluation of options for process sequences." Pp. 271–80 in *Proceedings of 13th IEEE Specialist Conference.*

Index

Basic research, xiv
Batteries, 56
Biomass conversion, 7–10
Building codes, 26

Capital, cost of, 163–65
Capitalization, 156
Clean Air Act, 150
Coal: costs of, 21, 150; environmental impacts of, 19, 139, 150; transition to, 134
Cogenerators, rights of, xvii, 114, 153
Combustion, 4–5
Conservation, 5
Consumer, 60–61
Crop residues, 8–9
Crude Oil Windfall Profits Tax Act (1980), 22, 52
Czochralski process, 54

Deforestation, 8
Delphi method, 51–53
Depreciation, 22–23
Discount rate, 21–22, 155, 160

Economic Recovery Act (1981), 88, 151

Economies of scale, 41–42, 52
Electricity: load cycles, 91–92; quality of, 113; replacement cost of, 22, 27
Electric Power Research Institute: on photovoltaics, 146; and synthetic utilities, 135
Electric utilities: capacity, 93–94, 101, 114–18; conventional performance, 94, 123–24; cost structure of, 74, 114–16; dispatching rules, 94–95; load-duration curve, 117, 121–22; and load factors, 106–8, 148, 150; loss-of-load probability, 103, 104, 125; rates, 22, 133–34, 166; reliability of, 103–4; securities of, 164; and tax structure, 146
Energy: cost of, xiv, 18–19, 151–52; demand simulation, 166–67; farming, 9; price of, xiii, 28, 140, 143, 165–68; problem dimensions, 1–5; risks of, 140
Energy Tax Act, 52

Fuels: as depletable capital, 165; external costs of, 98, 147; and industrial development, 2; risks from, 4; savings value of, 90–101

Gas: price of, 151; social cost of, 21
Genetic engineering, 9–10

Homes, passive-solar, 10–11
Hot-water systems, 29–36, 143–44; competitiveness of, 39; financing of, 26; and tax credits, 144
Hydroelectric resources, 142

Industry, research and development incentives, 152
Inflation, 159–60
Innovation: cost trends of, 50; inducement of, 44–46
Insolation: estimates of, 31; and irrigation, 68; and load patterns, 119–21; regional, 112
Interest rate: and consumption, 162–63; real, 23; in theory, 160
Investment: business perspective on, 22–23; and consumption, 39 n.2; criteria for, xv, 154–57; homeowners perspective on, 23–24; public-policy perspective on, 18–22; scenarios, 29–30
Irrigation, 65–67

Learning curve, 42–43, 50–51
Levelized costs, 34–36, 157

Mining, 4

National Photovoltaic Program, 57
Natural gas, 19
Natural Gas Policy Act, 52
Nuclear energy, 21

Oil: deregulation of, xiv; national-security costs of, 20; prices, xiii, 3, 19, 151; recoverable resources of, 2; social cost of, 21

Photovoltaics, xvi, 14–15; in agriculture, 65–71, 145; arrays output, 112; cell fabrication, 55; centralization versus decentralization, 112–14; costs of, 55–57, 147; development of, 47; effects of, 137–39; environmental advantages of, 139; fuel cell system, 56; fuel savings from, 127–33; homeowners' results from, 77–78; industry, 57–63, 151; as innovation, 64; as insurance, 133–34; learning curve for, 43; and levelized costs, 132; and load management, 118–23, 125–27; maintenance of, 124; manufacturing processes, 53–55; optimal purchase time, 71, 79–80; penetration in utilities, 78–79, 111, 122–23; profitability of, 67–71; progress function of, 48–50; prospects for, 145; from public-policy perspective, 75–77, 128–32; raw material for, 53–54; and reliability, 123–25; research on, 57; residential, 71–79; and substation transformers, 113; and transition to coal, 134–37
Plants, conversion efficiencies, 7
Power Plant and Industrial Fuel Use Act (1978), 114; and coal capacity, 117
Product development, 58–60
Public Utility Regulatory Policies Act (PURPA) of 1978, 27, 52; avoided cost pricing, 90; and time-of-day pricing, 56, 74

Renewable energy systems, 2, 24–29. *See also* Photovoltaics; Solar energy systems; Wind-energy conversion systems
Residential sector: energy consumption, 71–72; technical options for, 73–74
Resource exhaustion, 140–41
Risk, 29–36

Schumpeter, Joseph, 61
Semiconductors, 14
Silicon cells, 48
Solar energy: active systems, 11–12; avoided cost from, 148; and capital costs, 141; creative financing for, 157–59; economics of, xiv, 154; and energy demand, 141; environmental problems of, 142–43; and health risks, 16; as insurance, 36–38, 149, 153; investments, xvi, 26, 143; as manna, 5–6; market demography, 82; market prices of, 18; national-security benefits of, 151; as panacea, xiii–xiv, 15–16; passive

systems, 10–11; policy recommendations for, 151–53; and postpetroleum economy, 149–51; as process innovations, 17; scenarios, 30–32; as soft path, 147; and tax incentives, 152; as technological fix, 140; technologies of, 6–15, 142, 152; and utilities integration, 148–49; value of, 38, 143–45. *See also* Solar energy systems

Solar Energy Research Institute, 51

Solar energy systems: active, 11–12; and discount rates, 163; and existing technologies, 41–43; leasing of, 65; in multifuel energy systems, 111–12; net present value of, 32–34; passive, 10–11; prospects for, 46–47; resale value of, 39 n.3; risks of, 64

Sulfur dioxide emissions, 19–20

Sun rights, 25–26

Synfuels, 29

Tax credits: effect of, 146–47; and technological change paradox, 80–81; and wind energy, 88. *See also* Electric utilities; Solar energy

Technological change: paradox of, 60–62, 79–81; prospects for, 147–48

Technological forecasting, 47–57

Time-of-day rates, 113–14

Transmission, energy lost in, 112

Utilities. *See* Electric utilities

Waste, electricity from, 7, 8

WECS. *See* Wind-energy conversion systems

Western democracies, security of, 3, 140

Wind-energy conversion systems (WECS), 84; economic implications of, 88–100; and electricity prices, 90; fuel savings from, 95–96; 108–9; as insurance, 133–34; levelized cost of, 100; and load factors, 101–3, 106–8, 121; and loss-of-load probabilities, 105–6; performance of, 86–87, 92–93; potential for, 144–45; scientific principles of, 13–14, 86; and unreliability, 103–4

Wind machines, 85

Wind resources, 12–14, 85–87, 142

Wood, shortages, 7

SOLAR ECONOMICS

A user-friendly package of microcomputer programs designed for: investors in solar, wind, and other renewable energy projects, energy engineers, economists, and financial analysts, electric utility planners and consultants, public policy analysts and regulators.

This self-documenting package includes programs for:

- converting energy units from metric to English
- calculating the electrical output of photovoltaic and wind energy conversion systems
- measuring the avoided fuel and capacity costs from renewable energy systems
- calculating the present worth, payback period, and levelized costs of renewable energy.

Price: $175.
Program Diskettes available: Apple DOS 3.3; conversions for IBM Basic.

For orders and information, address the publisher
ROWMAN & ALLANHELD
81 Adams Drive   Totowa, New Jersey   07512